Wargaming in History

Series Editor
Stuart Asquith

Other titles in the series:

THE AMERICAN CIVIL WAR	*Paul Stevenson*
GOTHS, HUNS AND ROMANS	*Simon MacDowall*
THE SECOND ANGLO-BOER WAR	*Edwin Herbert*
WATERLOO	*Charles Grant*

The Peninsular War

•••••

Donald Featherstone

ARGUS BOOKS

Argus Books
Argus House
Boundary Way
Hemel Hempstead
Hertfordshire HP2 7ST
England

First published by Argus Books 1991

ISBN 1 85486 028 3

Phototypesetting by Photoprint, Torquay, Devon
Printed and bound in Great Britain by
Biddles Ltd, Guildford and King's Lynn

CONTENTS

INTRODUCTION

Napoleon is known to have called the Peninsular War 'the Spanish Ulcer' and there is little doubt that it played a substantial part in his downfall. More than that, it provided a host of relatively unchanged battlefields dotted around Portugal and Spain, each of which I have 'walked' at least twice during the past 25 years, savouring Wellington's sheer military genius and glowing with pride at the courage shown by our ancestors, the common British soldier of the Napoleonic Wars.

Writing this book on my 'favourite' war has revived many memories besides being a pleasurable chore, but requires a few

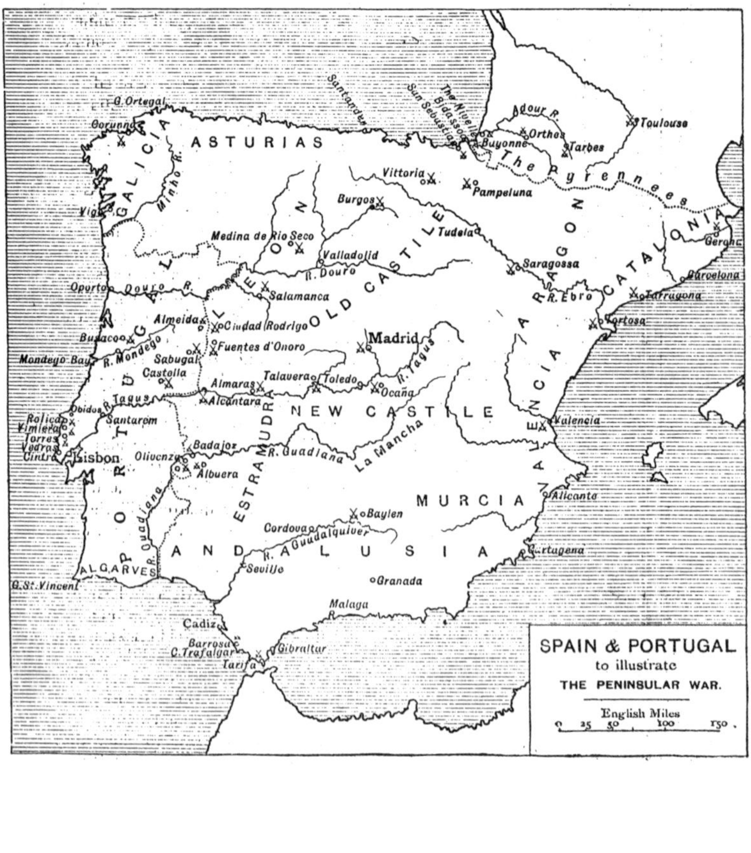

words of explanation. First, on the necessary condensation of battle accounts within the limitations of this book. Although this scarcity of detail could make for some interesting (if unhistorical) manoeuvrings on the wargames table, if authenticity is to be achieved, the wargamer/reader is recommended to study the highly readable and colourful accounts given in such readily accessible books as those by Weller or Glover (see Bibliography). From the same sources you will obtain more detailed Orders of Battle, as there is only room here for the actual number and formations of infantry, cavalry and artillery present at each battle. This book is about the British role in the Peninsular War, and the almost total absence of detail on the very significant part played by the Spanish Army is not intended in any deprecatory sense. You are recommended to read the highly detailed works of Esdaile and Gates (again, see Bibliography) if you have a large Spanish wargames army lying idle.

Worthy of the lyrical prose of Napier, the Peninsular War is perhaps the most satisfactory of all conflicts to wargame.

1 INTRODUCTION TO THE NAPOLEONIC ERA

During the Napoleonic era, the flintlock musket and bayonet, the smoothbore cannon and the employment of cavalry reached their peak, repeating and resembling the perfection of the Mongol and English tactical systems of the 13th and 14th centuries. Inevitably the full potential of gunpowder weapons would have been reached during the period leading up to 1850 but its advent was hastened by the genius of Napoleon Bonaparte, whose concept of warfare and his practical employment of men and weapons was to set standards for more than a century. His methods involved deception and rapid manoeuvre to outflank an enemy, cut his line of communication, and then force him to turn and fight at a disadvantage – as at Marengo, Ulm and Jena. Because the combined military strength of Napoleon's enemies invariably exceeded his own, it was necessary to deceive and confuse them by dissemination of his armies until the last possible moment, so allowing rapid manoeuvring and concentration, bringing superior force to bear at a critical point – Rivoli, Friedland and Dresden are typical examples. Or to throw a concentrated army between two hostile forces and defeat them in turn – by these methods Bonaparte was victorious in his first campaign at Montenotte, but, through an inadequate battlefield performance, failed at Waterloo.

Napoleon revitalised tactical methods, changing the characteristic European principles of formal linear tactics and sieges by rapidly manoeuvring so as to direct the main blow against the enemy's flank, perhaps simultaneously attacking his front or carrying out an enveloping movement against an enemy flank while sending the main thrust against his centre. Although contemporary artillery was neither capable of the rate of fire, or the accuracy or the trajectory necessary to closely support troops in attack, Napoleon's artillery reserve supported attacking divisions by massed fire.

The sheer fear-inspiring aspects of Napoleon himself must be taken into account – a famous commander of the time said that Napoleon's presence at a battlefield was worth 40,000 Frenchmen! As the years passed, the European commanders, learning from their defeats, imitated Napoleon's more obvious tactics; while this improved their performance, they never matched him. He was finally overwhelmed through numerical superiority and attrition.

Muskets and their ranges were much the same, the artillery equally effective and cavalry were similarly armed during the Napoleonic Wars as they were 50/60 years earlier in the Seven Years War and yet the warfare of the two periods was very different. When wargaming this is a fact that must be authentically reproduced by rules that are designed to emphasise and produce these differences. It is very easy to discover that one is merely fighting Napoleonic battles with armies wearing tricorne hats or a Seven Years War conflict with Napoleonic troops!

What are these differences and how does Napoleonic warfare compare with that of the 18th century? In this period Frederick the Great revealed himself to be one of history's outstanding commanders. Marlborough and Eugene were both exceptional soldiers; slightly lesser commanders displayed considerable skill – de Saxe, Villars, the Duke of Berwick, Vendome, Peter the Great of Russia and James Wolfe, to name a few. With the weapons still in use 100 years later, they all managed to win outstanding victories without availing themselves of such successful methods as column attack and their infantry frequently repelled cavalry without ever forming square! If these two tactics were seemingly universal and essential factors of the Napoleonic Wars, why did the acknowledged 'Great Captains' of the previous century not use them?

The simple answer is that they did not need to, they could win their battles without them! So, if the weapons were the same, then it must be that the soldiers themselves were different and that the rigorously disciplined and trained 'professional' armies of the 18th century were better soldiers than their counterparts of the early 19th century with the sole possible exception of Wellington's Peninsular army. And of course that was the case, as the vast armies of the Napoleonic period were composed of conscripts, with a small leavening of regular professional soldiers. These unwilling conscripts were often semi-trained, so that their musketry, of a far lower standard, could not bring decisions in firefights, and battles had to be won by attacking in massed columns in which the frightened levies were forced to go forward because they were pushed on by the pressure of their comrades behind and around them.

The cavalry of the Napoleonic period did not have that high degree of skill and manoeuvrability that marked Seydlitz's crack squadrons. They could only retain their formation and crash into the enemy as an organised body if they did not move so quickly, except over the last 50 yards of the charge. In the 18th century this would have given the infantry a great advantage but, by the time of the Napoleonic Wars, firepower was considerably lessened and infantry had to rely on the bayonet to hold back cavalry attacks. If the foot soldiers were in line formation, they could not stand if their flanks were threatened, and a cavalry breakthrough in any single part of the line could cause it to disintegrate. To prevent such a calamity, Napoleonic infantry battalions formed square which, although reducing their firepower even further as only a quarter of the formation could fire in one direction at the same time, was a strong and almost unbreakable formation that could not be outflanked. It also had the bonus of being, like the column, a formation that held the nervous conscripts together and prevented them from running away.

One of the most significant tactical developments of the period was the extensive emergence of the light infantryman – basically the French skirmisher and the British rifleman. Their true role was made possible by the introduction of the rifle, particularly needed by the light infantryman who had to fight as an individual in dispersed order but, because rifles were costly (and had a slower rate of fire), these specialised weapons were only issued to select units or individuals in line units.

It is a reasonable assessment of this period, which began in 1793 and ended in 1815, to say that the tactical methods of the 18th century were brought to fresh life rather than altered by Napoleon while he simultaneously adapted them to his own unique tactical and strategic concepts.

2 INFANTRY AND CAVALRY IN THE PENINSULAR WAR

The British sent arms, equipment and money to the Portuguese and Spanish people and an expeditionary force under Sir Arthur Wellesley landed in Portugal. Before taking the field, Wellesley reorganised his army by allocating a Portuguese infantry battalion, armed and organised in the British style, to each British infantry brigade, giving the local troops experienced backing so that by Busaco they had become first-class soldiers. Initially the British force was organised into eight brigades each of two or three battalions. Later it consisted of seven divisions, with a light division and cavalry under separate command. Usually British divisions consisted of about 6,000 men in two British brigades and one Portuguese brigade, each of three battalions, with a battalion of Cacadores (light infantry armed with rifles) to each Portuguese brigade. Every infantry brigade was given an extra company of light riflemen to reinforce the three light companies which were now standard in the British brigade.

The British army of this period was by no means typical, consisting of a small well-trained and disciplined professional force as opposed to the French and other Continental armies which were huge masses of citizen-soldier volunteers and conscripts. As a result, the British infantryman was more adept with his weapons, able to carry out more complex manoeuvres and withstand greater punishment than his Continental contemporaries.

An important part of the British force was the King's German Legion, formed from the remnants of the Hanoverian Electoral army, with a peak strength of two horse and four foot batteries; two regiments of dragoons; 2 regiments of hussars; two light and 10 line battalions each of ten companies of 110 men. The cavalry regiments consisted of four squadrons each of two 90-man troops. The King's German Legion light infantry companies had ten riflemen each so that their 100 per battalion was considerably more than a company per brigade. The artillery battery consisted

A model of a British infantryman posed against the gateway of a lovely piece of scenic terrain – a Spanish house of the Peninsular War period.

of six guns and 175 artillerymen, eight ammunition and two baggage wagons.

The cream of the British Peninsular army was the Light Division, with its essential role of countering the French Tirailleurs who skirmished ahead of their columns. Thoroughly trained, the British light infantry were able to operate in close order or as individuals in a manner far superior to the average French

Model soldiers posed against a beautifully constructed Spanish house of the Peninsular War period.

Tirailleur who frequently was only another conscript in a different uniform stuck out in front of the columns. A scattered rifle regiment could quickly reform to present a front to deliver the equivalent of musket volley. Impressed by the skirmishing tactics of the Colonists during the American Revolution, Sir John Moore had revolutionised British infantry tactics when he created the Light Division at Shorncliffe. Here Moore applied his theory that line and its firepower, with the two-rank formation allowing a much wider musketry front with the same number of troops, was the most effective defence against column attack. Realising the need to break up French shock attacks before they could be pushed home, Moore armed his Light Division with rifles, which doubled the distance and accuracy of their firepower. This weapon, coupled with first-class training in skirmishing, rapid movement and independent action made the soldiers of the Light Division the best infantry in the world.

In 1810, the British Light Division under General Robert Craufurd consisted of:

'A' Brigade – 1st Bn. 43rd; 4 companies 1st Bn. 95th and 1st Bn. Portuguese Cacadores.

'B' Brigade – 1/52nd; 4 companies of the 1/95th and the 3rd Cacadores.

Two battalions of the 95th, the 5th battalion of the 60th and the Cacadores of the Light Division were all equipped with the Baker rifle which could be loaded and fired, from a lying position if necessary, in 30 seconds but under battlefield conditions once per minute with an average misfire of 1 in every 13 rounds.

When in June 1809, Wellesley invaded Spain, he hoped to unite with some 100,000 Spanish Regular troops and guerillas; however, he soon realised they were poor allies. Throughout the Peninsular War, so far as Wellesley was concerned, with the exception of the experienced and battle-hardened men of Morillo's and Longa's units, he found Spanish troops to be unreliable and untrustworthy – possibly due more to the poor examples set by their commanders than to the men themselves. Generally speaking, their morale was unstable and, as exemplified at Talavera, after an opening volley that was misloaded, unco-ordinated and unaimed, there was usually a wholesale flight. Spanish cavalry would only go into action against disordered infantry; but their artillery, when not influenced by their countrymen fleeing all around them, frequently performed heroically and well.

A soldier of the Portuguese Infantry Regiment No. 9 – Division of the North; 3rd Brigade 1806.

An Infantry Musician of the 17th Portuguese Regiment (Divisao do Sul 2nd Bde.) 1808.

A Portuguese soldier of the Lusitanian Legion (Marqués de Alorna) 1808.

A soldier of the Infantry of the Portuguese Legion in the service of France 1808–1813 (Grenadier).

A cavalryman of the Portuguese Lusitanian Legion 1807–8.

A soldier of the 6th Portuguese Cacadores 1811.

A soldier of the 4th (Beira) Bn. of the Portuguese Cacadores – Companhia de Atiradores 1808.

An Artillery Officer of the Portuguese Division of the North; 18th Brigade (Braga-Vila de Conde) 1806.

Initially the French army in Spain consisted largely of recent conscripts, completing their training under service conditions. At Baylen in July 1808, an army of 20,000 of these raw troops stunned Europe by surrendering to a force of Spanish Regulars and guerillas.

In 1808 Napoleon had disposed throughout his various theatres of war, 113 regiments of the line and 32 light infantry regiments, so in theory there should have been 580 field battalions in the French army, excluding the Imperial Guard and various allies. In June 1808 there were, in fact, 417 field battalions on the books; and by 1813 there were 156 line regiments and 36 of light infantry. There was little, if any, difference in the training or tactics between line and light infantry. A field battalion had six companies of 140 men each – a total of 840 men in four fusilier companies and two flank companies. Regiments normally consisted of four field battalions and one depot battalion.

To enable wargamers to accurately scale-down their forces, the British organisation in the Peninsular War was:

INFANTRY:	50 men	=	1 company.
	10 companies	=	1 battalion (which included 1 Grenadier and 1 Light Company).
	2 Battalions	=	1 Regiment (Second Battalion usually in England).
	3 Battalions	=	1 Brigade (plus 1 Company of Rifles).
	3 Brigades	=	1 Division.

Thus, a battalion of ten companies will require 50 figures, including officers, colour party, drummers, etc. It can utilise the same shape trays, adapted for a two-deep formation of 6 men and 5 men – two companies with 1 officer and an NCO. Both light and grenadier companies skirmish independently so that when they retire into the square it will become more of an oblong – under battle conditions, infantry squares were not always geometrically precise! Both ranks will fire, so that the eight centre companies will bring 40 muskets to bear when in line or 44 for light regiments, against the French 9 or 18 muskets (double column) which is a very reasonable simulation of what actually occurred.

In this period infantry fought in the basic tactical unit of a brigade or regiment, formed of a varied number of battalions each consisting of about 600 men. In battle, the battalion formed up in three or four shoulder-to-shoulder ranks covering an area of about 200 yards which usually extended to as much as 400 yards because of the gaps caused by manoeuvring. To simulate this on the wargames table, battalions can be sub-divided into companies separated by small gaps representing the breaking of formations when marching over rough ground or passing obstacles. These gaps will permit the realistic simulation of parts of units being attacked by cavalry when beyond support of their comrades.

Throughout the Horse-and-Musket period, troops fought in tightly-knit tactical formations to ensure the steadiness and maintenance of morale that prevented them from being scattered or destroyed. If wargamers wish to reproduce the style of warfare of this period they must similarly organise their scaled-down numbers of figures into characteristic formations, while maintaining their specific grouping. Most wargamers have their figures in groups of three or four on a strip of card; however, moving any sort of a formation over a tabletop battlefield is a fiddling and time-consuming business that is made easier by having these strips on 'formation-shapes' – trays designed to simulate specific tactical formations. These shapes can be cut from adhesive-backed tiles that allow casualties to be removed and are cheap enough to be replaced after two or three battles.

The very nature of Napoleonic warfare makes it essential to have the correct tactical grouping of companies. For example, a French Napoleonic infantry battalion consisted of 4 fusilier companies; 1 grenadier company and a voltigeur company – each with an establishment of 140 men. To the scale of 1 figure = 10 men, a standard size tray is required that allows each fusilier company to be arranged in three ranks with 5 men in the front rank; 4 men in the middle rank covering the intervals and 3 men in the rear rank – all bases touching those of the men alongside them. The tray is cut so that its edges are square by the side of the front rank, tapering away to the width of the rear rank of three. One tray has a small square extended from its rear on which are stood the portes-aigles and the drummers.

The diamond edges of these squares allow them to be fitted together to form a square with a rectangular space in the middle for officers. The voltigeurs (on separate stands) will be deployed as skirmishers but can take shelter in the square, fitting into the corners which have been squared off. The grenadier companies were frequently drawn off and formed into separate battalions of grenadiers and they can be fitted onto trays to form such battalions. These trays permit all the standard manoeuvres of Napoleonic infantry such as deployment into line, formation of column, formation of squares etc.

In 1808 Napoleon had about 80 regiments of line cavalry; each had four squadrons of 150/200 men per squadron; regiments very rarely had more than 600 men. At that time, the breakdown was:

Heavy Cavalry	2 regts Carabiniers
	12 Cuirassiers
Medium	30 Dragoons
Light	26 Chasseurs-á-Cheval
	10 Hussars

Contemporary records indicate that, at various periods, the French also used in the Peninsular War the following types (no doubt embodied under the specific headings above) –
Grenadiers-á-Cheval
Gendarme's d'élite
Lancers
and, in the very early stages, a squadron of Mamelukes are mentioned!

British cavalry included –

Heavy Cavalry	Life Guards
	Horse Guards
	Dragoon Guards
Light Cavalry	Light Dragoons
	Hussars (KLG)
	Portuguese Line Cavalry

Wellington's entire cavalry force constituted a Cavalry Division, with three regiments to a brigade; three squadrons to a regiment (only two squadrons for Household Cavalry). Each squadron consisted of 120 men.

The cavalry was Wellesley's weakest arm and invariably he endeavoured to select battlefields that did not favour horsemen – fortunately the Spanish countryside made this possible. Outnumbered, it was not possible to match them against French horsemen in the big charges that the French were able to perform so expertly. British cavalry swords were made blunt by the metal scabbards in which they were encased whereas the French used wooden sheaths.

3 LIGHT INFANTRY IN NAPOLEONIC WARFARE

Light infantry had been introduced into European warfare in the 18th century when the Austro/Hungarian light troops frequently caused a lot of trouble to the armies of Frederick the Great. These 'specialist' infantry came into their own in North America, first during the Seven Years War and then in 1775 when the War for American Independence taught a bitter lesson to the close-packed ranks of British infantry. Their red coats had made them a striking target for the rifle-armed American frontiersman who struck with speed and accuracy while skilfully using all available cover. So great was their impact that the traditional conservatism of those in authority was shaken and, under the able direction of visionaries like Manningham and Moore, the British Army instituted light infantry regiments. Formed of men trained to act and think as individuals, clothed in dark green and armed with a weapon which out-ranged the conventional musket, these units played an outstanding and significant role in the Peninsula and at Waterloo.

It must be appreciated that there was a difference between the light and rifle battalions of the British Army in that the rifle battalions were armed with the Baker rifle which, although it took longer to load, had an effective range of 200 yards. The light infantry battalions had special smoothbore muskets capable of a sustained fire of four rounds in three minutes at a range of 120 yards.

In contrast, the French had light infantry 'thrust upon them' by force of circumstances. The volunteers of the early Revolutionary Wars lacked both the training and discipline to pursue formal drill movements and manoeuvres when under fire. Inevitably, they were repeatedly defeated until, in 1772, it became the practice to push forward the most reliable men in a cloud of skirmishers, so forming a screen behind which the nervous battalions huddled in fragile formations. When the skirmishers and artillery had shaken the enemy, the encouraged mass of infantry surged forward in wild bounds, often causing the astonished enemy to reel back. It was a tactic well suited to the native temperament and characteristics of the French Revolutionary soldier.

General Foy, a contemporary commander, described a typical engagement of this period:

'The action was opened by a cloud of sharpshooters, some mounted, some on foot, who were sent forward to carry out a general rather than a minutely-regulated mission; they proceeded to harass the enemy, escaping from his superior numbers by their mobility, from the effect of his cannon by their dispersal. They were constantly relieved to ensure that the fire did not slacken, and they also received considerable reinforcement to increase their overall effect.

'It was rare for an army to have placed its flanks in impregnable positions; in any case every position presents natural or contrived loopholes which favour an attacker. On such points the sharpshooters would concentrate their efforts, and élan and inspiration were rarely lacking at each time amongst such troops. Once the chink in the foe's armour had been revealed, it became the focal point for the main effort. The horse artillery would gallop up and open fire with canister at close range. The attacking force would meantime be moving up in the indicated direction, the infantry advancing in column (for it had little fire to offer), the cavalry in regiments or squadrons, ready to make its presence felt anywhere or everywhere as required. Then, when the hail of enemy

A British Rifleman of the 95th; Peninsular War period.

A French Cuirassier of the Peninsular War Period.

bullets or cannon balls began to slacken, an officer, common soldier, or, as often as not, a Representative of the People, would start to chant the "Hymn of Victory". The general would place his hat with its large tricolor cockade on the point of his sword so that it could be seen from afar and served as a rallying point for the gallant troops. The soldiers would begin to run forwards, those in the front ranks crossing their bayonets, as the drums beat the charge; the sky would ring to a thousand battle-cries constantly repeated: "En avant! En avant! Vive la Republique!"'

By 1795 all battalions were capable of acting as light infantry, dissolving into skirmisher swarms as soon as the battle began. These were not special light troops but sections of the regular infantry who had become more flexible. Later, light infantry battalions, such as the French Legère or British Rifles, usually sent forward no more than a half of their strength in skirmish order, with the remainder kept in formed order in the rear to serve as a reserve to support the forward elements, or to form a rallying point if the forward skirmishers were pushed back.

In 1805 there were 26 light regiments in the French army but by 1815, in spite of no less than 127 new regiments having been formed, the proportion of light units had diminished to about a sixth of the total infantry strength. During the three years that followed 1805, line battalions were reorganised and given six companies – one grenadier, four fusilier and one voltigeur (light infantry). The 1808 establishment gave the light infantry battalion one carabinier company (equivalent to line grenadiers); four chasseur companies (fusiliers) and one of voltigeurs or skirmishers.

One light battalion to four or five line infantry battalions is a reasonable proportion in a wargames army. Light infantry can be used as a screen for infantry attacks but are most valuable on defence as they can move and fire, firing then retreating etc. They can hover around infantry columns, harassing them with fire so that the infantry are finally forced to deploy to deal with them. Their open order skirmishing formation makes them very valuable in wooded or close country where a column or line of infantry are swiftly disordered. During the Napoleonic period, skirmish lines had one or two men per 5 yards of front; firing lines four to eight men per 5 yards of front, and battle lines (two or more lines) 16 or more men per 5 yards of front.

When wargaming, light infantry on the march will form up as a body like any other infantry unit and march at the same rate but when they operate individually as skirmishers they should be given a greater move-distance. This is counterbalanced by a loss of solidarity and cohesion because they must operate at least an inch apart whereas line infantry of this period fought 'base-to-base'. Light infantry operating less than an inch from their fellows will be treated as ordinary infantry and will move at the same speed.

One of the duties of the light infantry skirmisher was to pick off individual officers, and as most morale rules have a penalty for loss of officers, this can be a vital factor. A light infantryman can specify the officer at which he is aiming but it should not be certain that he actually hits him, a die is thrown and if it shows up 1 or 2 then it is the man immediately on the officer's right, 3 or 4 it is the officer himself, 5 or 6 it is the man on the officer's left who is hit.

In mêlées their properties are those of ordinary infantry although, when caught in the open by a formation of infantry or worse still by cavalry, their extended order makes them extremely vulnerable and they are very lucky not to be cut down. Rules should be slanted to cover these factors.

4 HANDLING NAPOLEONIC CAVALRY

Most European cavalry regiments of the Revolutionary and Napoleonic Wars were essentially alike, although there were differing minor details. First, there were the 'Heavies' – the shock-arm, partially-armoured men on big, powerful horses. The French, Austrians, Prussians and Russians knew them as Cuirassiers, while the French Grenadier-á-cheval and the British Life Guards and Horse Guards were also heavies. Their role was to batter down opposing enemy cavalry and to crush infantry masses, although these formations had first to be softened-up as even the heavy cavalry were relatively ineffective against fresh infantry given time to form square. Because they were highly vulnerable, heavy cavalry were rarely used early in the battle. French cavalrymen were generally better equipped than those of other nations – for example, their cuirassiers wore a back-and-breast plate, while the Austrians and most other countries gave their 'heavies' only a breastplate.

The Dragoons and the Carabiniers were medium/heavy cavalry, the latter (especially those of France and Prussia) had limited dismounted capabilities. Both Heavy and Medium cavalry regiments consisted of three or four squadrons, each of 150/200 men.

The Light cavalry consisted of many different types of mounted units – British light dragoons; French line chasseurs-á-cheval and Austrian chevauleger were equivalent types of units; hussars were used similarly and were perhaps a slightly superior type of cavalry. Presumably the Guard Light Horse, such as French chasseurs-á-cheval and the Mamelukes could be given a slightly higher rating although the fighting capabilities of all these cavalry were much the same. Lancers were used in massed formations for flank protection, attacking infantry (especially when they were in square) and supporting heavy cavalry. The Austrians had lancers; the Prussians had the Uhlans and the French had both the Chevauleger Lancier and the Polish Lancers. The Russians Cossacks were not really battlefield cavalry but lancers used in the same way as the British Light Dragoon – light cavalry

An Officer (Lieut.-Col.) of the 14th Light Dragoons in Parade Dress.

employed on scouting duties. The French equivalent was the Scout Lancer of the Guard. The British had no lancers.

'Lights' on smaller mounts were used for harassing, reconnaissance and shock tactics, being most effective *en masse* and in conjunction with Horse artillery, particularly when thrown against infantry previously shaken or shattered by artillery fire or infantry attack. Line Light Cavalry regiments consisted of 600–1000 troopers, but some had as many as 1,500 sabres.

Led by outstanding leaders, Napoleon's cavalry benefited from the French temperament and were usually superior to the best cavalry of other countries who themselves steadily improved their cavalry by copying French organisation, tactics and methods. Wellington's cavalry played a minor role in the Peninsular Campaign but at Waterloo they smashed a final attack of the Old Guard.

It is possible that Napoleonic horsemen did not have the same proficiency and well-drilled manoeuvrability of the highly professional Prussian squadrons of Seydlitz and, to retain their formation and charge home as an organised body, they had to reduce the speed of their charge until the last few yards. Cavalry at this period could charge at the rate of about 700 yards per minute and their charges were carefully graduated – the squadrons trotted the first third of the charge-distance, cantered until a gallop was attained at 150 yards from the enemy then, over the last 50 yards the horses were allowed their heads and flat-out charge-speed was reached. On the battlefield this required high-class control and calculation by experienced leaders; much the same qualities must be written into wargames rules that control simulation of the Napoleonic cavalry charge.

Cavalry charging formed infantry would have to traverse 350

A Private of the 13th Light Dragoons.

French cavalry of the Peninsular War.

yards of fire at the canter; then 100 yards at the gallop and finally, 50 yards at actual charge speed. But the limitations of the firearms of the period caused only the last 150 yards to be particularly dangerous. Thus, the limitations of Napoleonic cavalry were balanced by the relatively decreased fire-effect of infantry of this period, and the infantry were forced to rely on the bayonet to hold back cavalry who could do little against steady infantry in a defensive formation which allowed them to fire volleys and present the bayonet. When infantry were charged, the second line of the cavalry was only 200 yards behind the first so that a battalion which had expended its musketry against the first line would not be prepared for a second charge coming in so quickly. At other times, when cavalry charged in two lines, the second line was 400/500 yards from the first so that it could take advantage of any success gained by the first line. It was sufficient distance for a defeated front line to pass back without disordering the second line.

So, cavalry engaged enemy cavalry when there were no disorganised infantry formations to attack. In these cavalry mêlées, victory usually went to the most disciplined force who held their reserve until the appropriate moment. The formation of a cavalry squadron was soon dispersed in an action and self-preservation usually caused an officer to be followed whereas the trooper of a broken cavalry formation tended to seek safety. Realistic rules should penalise both the fighting power and the morale of a cavalryman once he has lost his formation.

The French Marshal Marmont believed six squadrons to be the maximum possible frontage, with the first line several squadrons deep and flanked by two columns of squadrons capable of forming front to a flank if threatened and able to peel off and attack any unprotected flanks presented to them. The second line (in column of squadrons) remained some 400 yards back, acting as a reserve and a support, spaced to allow them to wheel right or left or to form line.

At scales of 1 figure = 20 men, a cavalry squadron will consist of 6/7 figures and, even with only six squadrons, it is possible to simulate the actual formations of the Napoleonic Wars. The wargamer might be reluctant to put only half of his horsemen in the front line against an enemy who might overwhelm them by using all his troops simultaneously. But only the actual frontages (plus a small 'overlap') will count in the first round of the suggested mêlée-method and, as the larger line has been disordered they will act at a disadvantage against the formed and controlled second line.

A squadron was usually formed in 50 files of one front and one rear rank man, producing a two-deep line on a 50-yard frontage. A squadron's strength varied according to armies and to circumstances – British squadrons were usually about 100–150 men; French about 120–150; Prussian about 90 men. Three or four squadrons formed a regiment although some Prussian Light Horse regiments had five squadrons.

5 FRENCH COLUMN AND BRITISH LINE

Even before the Revolutionary Wars, France believed that shock action should be delivered by formations with a greater depth than the firing line and that supporting fire should be delivered by swarms of skirmishers. The French attack column, introduced in the Wars of the Revolution and developed by Carnot, was forced upon the commanders of the day as the best possible way to use the ill-trained and badly disciplined troops produced by the *levée-en-masse*. The physical advantages of forming a massive column of a number of battalions were enhanced by an undoubted morale effect both on troops forming the column and on the enemy towards whom it was directed. Possessing the tactical assets of flexibility and versatility, it permitted the rapid and controlled movement of large numbers of men in the face of the enemy. Able to be used in hilly country, a column could quickly take up different formations such as two/three rank firing-lines and squares; skirmishers could be detached and in conjunction with the artillery, inflict casualties on the enemy.

Napoleon perfected this system by employing his infantry in a derivation of Frederick the Great's line, broken up into highly mobile parts through battalion columns of companies or double companies, their fronts covered by skirmishers.

Frederick's system was to fire first and then charge but Napoleon's infantry did not attack until skirmishers had cleared their front and their artillery had shaken the defenders. Obviously, much depended upon the volume and affect of the preliminary fire, which Napoleon later increased by supplementing his skirmishers with massed artillery firing grape or case shot.

In the later years of the conflict, Napoleon found it increasingly difficult, because of lack of experienced officers, NCOs and men, to utilise even relatively sophisticated formations at a regimental level. Perforce, he employed increasingly larger formations until, at Wagram in 1809, a frontal attack was made by 28 battalions formed into a single huge column. By 1815, it was common for

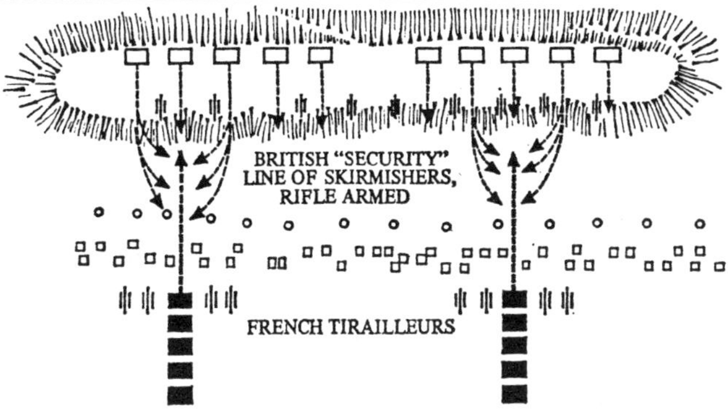

Wellington's Defensive Position. British battalion columns at quarter distance on the reverse slope. On forward slope or crest British light artillery are positioned. This schematic representation of Wellington's counter-tactics is reproduced from *Weapons and Tactics' Hastings to Berlin* by kind permission of the author, Jac Weller, and the publishers Nicholas Vane.

Wellington's Defensive Position in the Peninsular War.

complete divisions to manoeuvre as a large column. Such massive formations presented unmissable targets and suffered heavy losses, although Napoleon may well have considered them justified if a breakthrough was achieved.

It was as a 'sustaining' force that the French attack column was most frequently employed, serving as a replacement pool for the outlying skirmish line. Having little firepower when on the move, once the dense formation reached the enemy the column could deploy into lines for a firefight if firm resistance was encountered. In their attack on the Pratzen at Austerlitz, the troops of Soult and Bernadotte approached in column, deployed, fired and reformed column for the charge. Once the enemy showed signs of wavering, the lines could advance or reduce their front and move forward again in column. Frequently, this brought success because, if an infantry formation had been distracted or disordered by being shelled and harassed by skirmishers, it only required an almost nominal forward movement by attackers to disturb the formation so that it probably moved away. It was extremely rare for bayonet attacks to be driven home against a steady enemy and hand-to-hand fighting was most infrequent as one side usually gave way before bayonets were actually crossed. (Napoleonic mêlées on a wargames table should rarely last more than one move.)

It was Wellington and his small British Army who supplied the most effective solution to the otherwise successful French tactics. Involving the best practical use of light infantry, the British method employed maximum firepower, in the form of controlled and aimed musketry, delivered by well disciplined troops who combined the mobility of skirmishers with the steadiness of line infantry. In the face of the acknowledged success of the French column, the British retained the two-deep line, usually taking advantage of cover until they rose to pour into the column a far greater and more devastating volume of fire than it could return. Then, with the column in some disorder, the British line folded around the head of the columns in a bayonet attack that could not be combated by its closely-packed inner ranks. Prior to receiving the infantry fire, the column and its accompanying artillery had been galled and harassed by the well-aimed fire of the screen of riflemen thrown out in front of their line.

Only steady disciplined troops could operate in the British manner, and Blücher remarked that his troops, even in three-deep lines, did not possess the discipline, training and individual stability to hold such formations, so they did not use them. The British and eventually their Portuguese allies in the Peninsular War were professionals with superior qualities to the conscript – this enabled them to deploy in line and so employ maximum firepower.

French columns often outnumbered British lines but were invariably repulsed in a manner that astonished the French, aware that their massed formations, tirailleurs and artillery had defeated the best that Europe could throw against them. The musketry advantage that the infantry line had over a column was usually outweighed because they were so weakened by the fire of light infantry and close-range artillery that they were overcome by the shock effect of the columns. This was all altered by Wellesley's feat of countering the tirailleurs with his own riflemen while the artillery accompanying the columns were unable to deliver supporting fire because horses and artillerymen were casualties before they could come into action. Like King Edward, the Black Prince and Henry V during the Hundred Years War, before he would give battle Wellesley invariably sought a position, preferably on back-slopes, that best suited his troops. The greatest single virtue of these defensive positions was that they always permitted a rapid and continuous flow of reinforcement to threatened sections, often on quickly constructed roads running behind the position.

An assault with more than one battalion in column is best delivered with a frontage of two companies, whether the battalion advances parallel or in succession. The reason for this is because, at full strength, a one-company front will be 40 men wide by 25 men deep (excluding officers, eagle, etc.). Using trays, the column will be 24 men deep but will only have a frontage of 5 men. If the suggested organisation of 72 other ranks per battalion is too high for the wargamer's available resources, it can be cut to 60 if the grenadiers are separately organised or 48 if separate voltigeurs are provided. However, if smaller battalions are used then the three-deep formation can be reduced to two-deep companies of either 9 or 7 figures.

When French columns came under fire from British lines they were forced to abandon their attack in the endeavour to hold their ground. With time to adjust and convert the action into a static firefight (as at Albuera), the resulting action was in the 18th-century style, usually lasting about 20 minutes before one side broke, often because one of them showed signs of coming forward, as when reinforced. British infantry attacks on columns were made immediately after their first volleys, which were rarely the oft-mentioned 'rolling half company volleys for long periods', because the musketry was a preliminary to the possible use of the bayonet which, however, was rarely used as formations broke and fled before contact. When wargaming, there should be more stress on morale than on formation in this aspect, although formations were important in firefights. The British system of halting a column attack with volley-fire and then immediately charging the demoralised survivors can be simulated by splitting each move into two or three parts so that one player takes a 'moving' move while his opponent takes a simultaneous 'firing' move. The 'moving' move is a three-part move and allows charges etc, while the 'firing' move is a two-part move that permits the unit to wheel or change formation and then fire. Each unit is taken in turn and the choice of the unit can be made when its turn arrives.

In the Peninsula, French column attacks failed because they were made before the British were disordered, without the French realising this fact. When the fire of the skirmishers failed to disorder the British, then the attacking infantry column sought to fire to attain this effect but French attempts to deploy against the British were usually too late and resulted in confusion.

Line formation possessed a very marked superiority in fire-power over the column, which presented a far larger target. As an example, taking up-to-establishment battalions more than 1,000 strong, the two-deep line covered an area of about 316 yards by 1 yard and could bring *all* its muskets to bear, less four

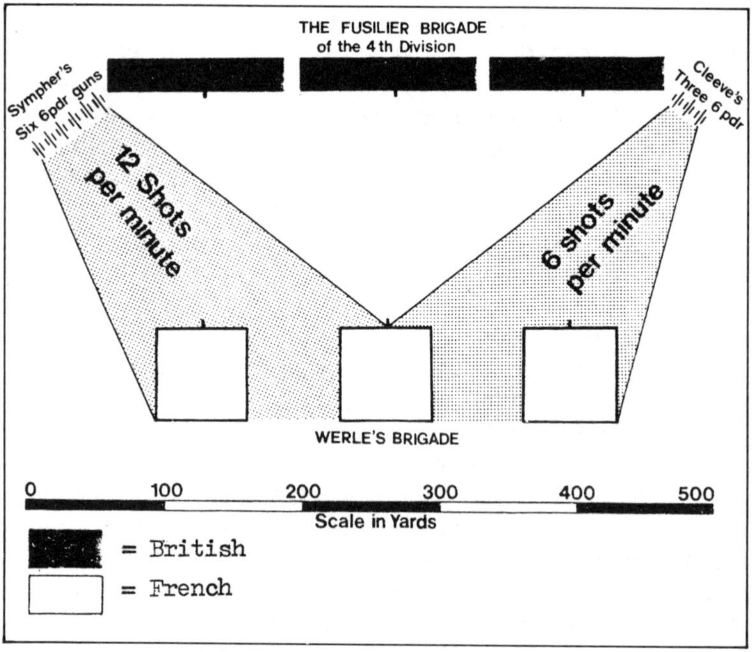

Column v Line. An example (based on what occurred at Albuhera) of the fire that could be brought to bear on French columns.

which bore to the flank. Whereas the column, in nine ranks covering an area 75 yards by 15 yards, could only bring about 248 muskets to bear frontally and 18 to the flank.

A WARGAMING METHOD OF SIMULATING INFANTRY ATTACKING IN COLUMN

A column may be:

a) A battalion column of 4 men wide by 10 men deep.
b) A regimental column of 4 men wide by 20 men deep.

Alternative columns may be formed of any number of men wide providing that the depth of the column contains more figures than its frontage.

An attacking column may fire, with appropriate penalties for moving, during its approach although the only people capable of doing this are those in the front rank and on the outside of the files. Should a column lose one-third or more of its strength in the approach it must test its morale or dice to see whether it charges home (I suggest 4, 5 or 6 to do so). Otherwise the column turns at the halfway point and retreats back to its starting point. It is probably more authentic to so weight morale that a line does not stand against a column unless they are very firm and of high quality. On the approach of the column, the defenders must dice to see if they stand and face it (this can be done at the moment when the column reaches its halfway point). If their morale is not high enough to stand then they retreat a specified distance in good order (or in disorder) according to their morale rating. When the column reaches the enemy, it drives in 3 inches, pushing before it the same number of men (regardless of the number of ranks) as form the front of the column i.e. a column 4 men wide will push a gap 4 men wide in the enemy line which may also be 1, 2 or even 3 ranks deep. The first round of the mêlée takes place between the front rank of the column and the front rank of the men it has pushed back, with those in the column being given some sort of bonus to represent the shock of impact. While this is going on, the remainder of the defenders have to curve back to conform and in the second round of fighting those defenders who are in contact with the flanks of the column take part against those flanks, who also get a bonus for shock effect.

Unless the defenders push the column back after both rounds of fighting, the column is considered to have broken through the line and can move its full distance with remnants of line brushed aside on its flanks.

The Duke of Wellington had no misconceptions about column and line. In 1811 after the action at Sabugal where four British battalions had repulsed 6,000 French infantry supported by cavalry, he wrote 'Really these attacks in columns against our line are very contemptible'. A few days later, he wrote 'I do not desire better sport than to meet one of their columns *en masse* with our own line'.

6 'FORM SQUARE!'

Heard over the noise and tumult of battle, this order must have aroused conflicting emotions in the breast of the infantryman who, while welcoming the comfortable feeling of security arising from being tightly surrounded by his comrades, was aware that it betokened a cavalry attack or hails of grape shot whistling through the packed ranks.

An infantry square was usually formed by a battalion although two battalions sometimes combined; in Egypt and Russia the French divisional squares had artillery at the corners and between the battalions – cavalry, headquarters and train sheltered inside the formation. Whenever possible, squares were positioned in echelon and spaced so as to be capable of mutual fire support without endangering each other. From column of companies it took about 30 seconds to form a square; wargames rules should allow for a time lag to cover awareness of the need to form square and for the necessary orders to be given. The formation was taken up in the following manner:

With battalion in line, the two centre companies stood fast and the remainder wheeled or filed back to form the flank and rear faces of the square. **From open column**, forming on the leading company, the centre companies wheeled to right and left to form the flanks while the rear company closed up and faced about. **From close column**, the rear company faced about; filed on the flank of each centre company faced outwards with NCOs filling up the small gaps between the flanks of the companies.

British battalions formed square with a front face of four ranks each of 34 men and rear faces of four ranks each of 26 men. Each face was about 60 feet in length and the open space inside was about 39 feet wide. The front ranks knelt and, after firing the first volley, 'set' their bayonet-tipped muskets in the same way as pikes; after firing their volley, the second ranks crouched low, the third and subsequent rank did not fix bayonets. All other European armies normally formed six ranks deep on each face although there is considerable recorded evidence that indicates three ranks.

It is not unreasonable to extend to the British infantryman of the period a greater degree of resolution (i.e. higher morale status) than the majority of his Continental counterparts because Wellington's infantry invariably remained where they were placed in square and brought down any horse or man who came within contact. In square formation British infantry were almost unbreakable; at Fuentes d'Onoro in 1811, the Light Division, formed in five squares, were attacked on an open plain by four brigades of cavalry and yet had only 35 casualties while retreating two miles!

Accurate musketry fire from steady lines of British infantry are claimed to have 'usually repulsed a cavalry charge' (Dupuy *Encyclopaedia of Military History* p. 735). Presumably this was in the Peninsula but the actual instances are not cited. Certainly, at Albuera in 1811, Napoleonic lancers caused 75 percent casualties in three British Line battalions in three minutes but this was due to the infantry, in open-flank line formation, being taken in flank and rear rather than to the rainstorm which reputedly prevented their muskets from firing. In the Napoleonic Wars, if an infantry battalion was attacked by cavalry while in line or while forming square it would be destroyed, and once infantry were broken they could be ridden down and annihilated. This happened at Quatre Bras where the Prince of Orange lost one battalion and then two more at Waterloo, besides damaging four others, by giving incorrect orders so that they were caught by cavalry while not in square.

When attacking a line, cavalry were forced into its centre because there was nowhere to swerve to avoid impact and, providing the line did not look too solid, the horses would go forward. The 3-deep line was the very minimum depth that allowed the slightest chance of warding off cavalry, but it could always be overlapped and its flanks turned except as at Salamanca, where Ferrey deployed seven infantry battalions in a 3-deep line with a battalion in square on each flank.

A combination of cavalry, artillery and infantry were most likely to succeed against a square, which was a packed, almost unmissable, target for artillery and musketry while its own firepower was reduced by three-quarters as only a quarter of the men facing the enemy could fire. Because it could always outrange them, canister against squares was a classic tactic – Wertingen in 1805 is an example of this, while at Quatre Bras, steady infantry (the 42nd Regiment) caught in the act of forming square, were able to beat off cavalry who had outrun their guns.

With the cavalry close enough to force the infantry into square, horse artillery could flay its packed ranks with canister; then supporting infantry moved up and deployed into line (the infantry in square dare not deploy for the forthcoming firefight for fear of being overrun by the nearby cavalry). With an advantage of 4 to 1, the attacker's fire could so shake the square that the supporting infantry were able to charge home.

To prevent this happening, enemy horsemen were held off by a friendly cavalry force and the square supported by horse artillery thrown well forward or in the space beween the squares. These guns fired until the last moment then the gunners ran to the protection of the nearest square, throwing themselves on the ground beneath the front-rank bayonets. If time allowed, they removed a gun wheel and brought it back with them. Courageous gunners like Mercer's men at Waterloo, remained with their guns, sheltering beneath them while the cavalry attack flowed past on either side.

One of the few occasions in military history when cavalry have broken infantry squares, drawn up on ground favourable to the infantry, occurred at Garcia Hernandez during the Peninsular War. As they advanced in echelon to charge French cavalry, Van Bock's Dragoons of the King's German Legion came under enfilading musketry fire from French infantry drawn up in square. The action is stirringly described by Major-General James Lunt, a former commander of the 16th/5th Lancers (*Charge to Glory* by James Lunt, 1961 pp. 82/3).

'Captain Von der Decken, commanding the left squadron, could see the effect of this fire on the leading squadron, and he knew that he must run the gauntlet of the same death-dealing musketry. He at once decided to charge the nearest French square, although the ground favoured infantry more than cavalry and despite the fact that there was no artillery available to redress the balance and clear a passage for his sabres. Wheeling his squadron into line, he led it headlong against the French, who greeted him with volley after volley. The dragoons enveloped two sides of the square but could not break into it. The two front ranks of the infantry were kneeling, and behind them were four more ranks standing. Fluttering over the centre of the square were the bullet-torn colours, and out in front were the lightly equipped skirmishers picking off the leading dragoons. Von der Decken was one of the first of his squadron to fall.

'In face of such a wall of musketry and steel, it seemed that the wave of horsemen must break and fall back, but at the crucial moment a shot killed one of the dragoon horses and it fell with its rider on top of the French ranks. The opening was made and with a wild yell the dragoons urged their terrified and snorting horses into the gap. It was like the flood which follows the bursting of a dam. The infantry disappeared under the onrush of slashing, yelling, thrusting horsemen, the confusion made a thousand times worse by the plunging and kicking of the fear-maddened horses. By the time the dragoons were finished the entire battalion had either been cut down or taken as prisoners, and where the square had once stood lay the disembowelled carcasses of nearly forty horses.

'Meanwhile the third squadron of the 1st Dragoons under Captain Von Reitzeuslen had followed Von der Decken's example and charged the second French square. This was drawn up on the edge of the heights and it was even better placed to drive back cavalry. But its morale was shaken by the fate of the other battalion. The cohesion essential if infantry was to withstand the shock action of cavalry was ruined by a number of soldiers who broke the ranks and took to their heels. The dragoons smashed their way into the gaps left by the fugitives and sabred their way through to the other side of the square; then, rallying as one man to the trumpet-call, they rode back the way they had come and completed the rout.

'A few unwounded French officers hastily rallied the survivors and attempted to form a third square. This in its turn was charged and destroyed by the two squadrons of the 2nd Dragoons which had been following behind their sister regiment. Foy's rear-guard had ceased to exist as an effective fighting force, and all who remained alive and untaken were withdrawn to some rising ground near the Peneranda Road. The King's German Legion followed them, but by now their horses had been virtually ridden into the ground. The French held them off with the courage that is lent by despair, picking up the stones which littered the ground and hurling them at the legs of the heaving and staggering horses. Wellington then sent orders for the dragoons to retire, and they made their way slowly back to the rear through the cheering ranks of the British infantry. They left behind them in the valley of Garcia Hernandez fifty-two dead dragoons and sixty-seven dead horses; their wounded totalled slighty more.'

Wargamers have an unfortunate propensity for seizing upon such isolated incidents as Garcia Hernandez and the breaking of the Sikh squares by the 16th Lancers at Aliwal as valid reasons for formulating rules that ignore historical reality by giving far too much power to cavalry and far too little to infantry in square. On the very rare occasions that cavalry broke into infantry squares, the foot soldiers were exhausted, their morale was low or they had taken heavy losses; rain rendered muskets useless and lances sometimes prised an opening. During the Napoleonic Wars the lance was an effective weapon against the sword or bayonet, but, when used against trained infantry in square who could fire musketry volleys, lancers do not appear to have been any more successful than other types of cavalry. And it was the presence of artillery rather than a cavalry threat that forced a square to surrender at Dresden.

Rules for wargames covering Napoleonic warfare must embody conditions that prevent cavalry from defeating steady infantry – a square with reasonable morale should be almost invulnerable to a cavalry charge. History tells us that in every battle (Jena, Wagram, Dresden, Ligny and partially at Quatre Bras) where the French 'armoured battering ram' was successful, the infantry were already exhausted or unbalanced. It is recorded that, on many Napoleonic battlefields, low-morale infantry formations broke before forming square at the terror-inspiring sight of the French heavy cavalry thundering down on them.

Rules must reflect the fact that when cavalry attack an infantry square it is a psychological rather than a physical battle. As the cavalry get close then the infantry, if they are not steady, may break, otherwise the cavalry will react to the steady lines of gleaming bayonets held by resolute infantry whose fire-discipline has already enabled them to take toll of the onrushing horsemen. Then, the cavalry will refuse to go into the bayonets and either stop or swerve to pass along the side faces of the square.

The length of the face of a square only allowed about 18/20 cavalry to attack it at a time; being in more than two ranks added little if any power to the attack and bodies of men and horses piled up in front of the square from previous attacks gave it increased strength and protection. Horses dislike stepping on a prone man and refuse to charge into a solid obstacle so that a bayonet-fringed, solid-looking wall of unyielding infantry caused cavalry to swerve round the 'island'.

Rarely, if ever, did the cavalry actually charge home on a square – Sgt.-Major Cotton, a Waterloo veteran, recorded that not one of the 15,000 French cavalry perished on a British bayonet nor did any infantry in square fall to a cavalry sabre.

Thus, wargames rules must give the cavalry a chance by testing morale to see if the infantry stand in the face of a charge, which will be carried through if the infantry falter. Remember, if a single horseman penetrates the square then others can follow!

It was a question of morale, balanced between big men wearing cuirasses and riding large horses against an unflinching wall of infantry fringed with 18-inch bayonets. The deciding factor was the coolness of the infantry which obviously affected the quality of their musketry. Accurate musketry required steady infantry who could wait until the enemy was right upon them before they fired – at Quatre Bras it 'broke and shattered' the French cavalry. Even when fire was delivered at 30 paces, Fortescue relates that cavalry was turned back in confusion. The number of casualties inflicted upon the cavalry was not always decisive (the Thin Red Line at Balaclava killed few Russian hussars but turned them back); there are recorded instances of battalion volleys at 100 yards only hitting four men. When charged, infantry cannot be expected to kill as many cavalry as they could infantry and rely upon artillery to support them in this.

When in square, firm infantry such as the British relied mainly on the firepower of their muskets and it was bullets, not bayonets, that caused enemy casualties. As a general rule, the infantry could only expect to get off one round before their range was down to bayonet length. Suvorov claimed 60 paces as the maximum effective range while, at the Battle of the Pyramids, French squares held their fire until the Mamelukes were within 50 paces.

When cavalry are within charging distance of a square, then those infantrymen are foolish to fire until the charge actually takes place because it is impossible to break a square of steady infantrymen with loaded weapons. If the weapons are unloaded then there is a slight chance of disaster although the cavalry were halted by the unyielding hedge of bayonets plus the noise and the fear which increased the horseman's impression of the steadiness of the infantry formation.

At Waterloo, many infantry battalions did not fire their muskets for long periods and Wellington is quoted as saying 'Our people would not throw away their fire till the Cuirassiers charge, and they would not charge until we had thrown away our fire. A quarter of an hour . . . passed without firing. It ended by our Squares advancing and driving off the Cuirassiers.'

7 ARTILLERY AND SMALL-ARMS

The Gribeauval field-artillery system employed by the French exemplified mobility, achieved by lightening the weight of both barrel and gun-carriage, so that Napoleon's artillery arm required only six horses, disposed in double-file to draw a 12-pounder gun, while four draft-horses could easily haul 8 and 4-pounders and the 6-inch howitzer. However, the bad roads of Portugal and Spain caused grave transport difficulties, plus on occasions a shortage of horses – at Fuentes d'Onoro, Massena had borrowed many teams from Bessieres. For these reasons the proportion of guns to men was much less in the Peninsula than in the Grand Army, as the following table shows.

Commander	Battle or Campaign	Date	Guns	Number of Men	Approx. Proportion of men to guns
Junot	Vimiero	1808	23	13,500	580
Victor	Talavera	1809	80	50,000	620
Massena	Portugal	1810	126	70,000	560
Massena	Fuentes d'Onoro	1811	42	40,000	1000
Soult	Albuera	1811	c.40	24,000	600
Napoleon	Russia	1812	1372	600,000	440
Napoleon	Germany	1813	1056	450,000	430
Soult	Pyrenees	Autumn 1813	125	107,000	860
Napoleon	Waterloo	1815	350	120,000	340

The largest French gun was the 12-pounder but it was seldom used; it threw an iron ball, 4.75 inches in diameter, on a flat trajectory so that it could harm only those directly in its path as it did not explode. The smaller guns threw smaller balls, but whatever their size a ball could not hit a man if he was positioned behind a ridge or in a hollow such as a sunken road. A 12-pdr shot given 6° of elevation could carry 1900 yards; the smaller 8

9-pdr gun and crew of Royal Horse Artillery. Peninsular War period.

and 4-pdrs could carry almost as far, but it was considered that beyond 1100 yards objects became too indistinct for accurate fire. The maximum range of artillery was calculated to the point where the roundshot pitched when fired at maximum elevation, beyond pitching-point a shot would ricochet on dry ground for a mile or so, to lethal effect. Artillery was ideally sited in such a way that a grazing-fire was achieved; a slight elevation was preferable to a big one, as this produced plunging-fire, causing the ball to bury itself and greatly reducing its effect. Howitzers had a range of about 1,000 yards at 4° elevation, which provided four or five ricochets on favourable ground. It is significant that effective-range was considered to be rather more than half maximum-range: French theorists of the day considered 1100 yards as the maximum effective-range in war.

It might be of value to wargamers to note that eight men could handle a French 4-pdr; eleven were needed to move an 8-pdr; and fifteen to manhandle a 12-pdr.

The French system of fighting employed irregular swarms of skirmishers covering their front to shield the column behind them and disorganise the enemy line by their irregular fire. Napoleon increased their effectiveness by adding field artillery, going forward beside, or even in front of, the infantry columns, opening fire at the effective canister-range of about 300 yards. Napoleon and his commanders realised the value of concentrated fire on a single point from a number of angles, consistently using their artillery effectively in both offensive and defensive roles.

Having no liking for counter-battery fire, Wellington preferred to use all artillery firepower to turn back infantry masses, employing his guns selectively in small numbers and individual batteries carefully sited all along his front as a support for his infantry and as an essential part of his defensive/offensive tactics. Field and horse artillery batteries, then called brigades, usually consisted of five 6-pounder guns and one 5.5-inch howitzer, although both 3 and 9-pounders were also used. The proportion of guns to infantry was traditionally low, being about two per 1,000 bayonets. The gunners were extremely well-drilled within their batteries and, being selected for size, were capable of manhandling their pieces across rough terrain and into inaccessible places.

The wargamer might find it useful to know the organisation of a troop of horse artillery equipped with six 6-pdrs which required 5 officers and 155 NCOs and men, including the drivers. A field-artillery battery, equipped with six 9-pdrs required 163 officers and men, plus 116 horses in gun-teams of eight horses. The horse-artillery troop had 156 horses; and also required six ammunition-wagons (one per gun) each drawn by six horses, a forge, a spare gun-carriage, and a store-wagon, the only four-horse vehicle in the troop. An artillery-wagon carried about 100 rounds and it is thought that at Waterloo British artillery expended 9,467 rounds, requiring a lot of transport! At Quatre Bras and Waterloo, Wellington's army expended 987,000 musket-cartridges, or about a hundred cart-loads.

The British occasionally used rockets in the Peninsula, de-

A Rocket Troop. Figures and construction by Stuart Asquith.

Royal Artillery of the Mounted Rocket Corps
The rocket shell, or rocket as it was called, was developed by Colonel W. Congreve and produced at the beginning of the 19th century.

Early in 1811, a detachment of Horse Artillery was placed at Congreve's disposal and during the next two years he continued his experiment and gave demonstrations on Woolwich Common. In January 1813, 194 officers, N.C.O.s and men were added to the strength to act as Rocket Detachments; and also a depot supply for troops on service. In 1814, the various Rocket Detachments were combined, those at home becoming the 1st and those abroad the 2nd Rocket Troop.

On the 13th October 1813 during the Battle of Leipzig the rocket battery fought so effectively that the Prince Regent insisted on Wellington accepting the Rocket Troop for service with the Peninsular Army. During the Battle of Waterloo, they fought alongside the Royal Horse Artillery and were heavily engaged.

The Rocket Troops were disbanded after 1815.

signed by Sir William Congreve, to be used as an intermediate-range weapon bridging the gap between the musket and artillery. However, their notorious inaccuracy and limited maximum-range of about 1,500 yards made them unpopular. They were hated by Wellington who said of them: 'I do not want to set fire to any town, and I know of no other use for rockets.'

Other than roundshot, explosive projectiles known as 'common shell' were used, being capable of bursting in the air, over the heads of troops; it was useful for searching behind crests at ranges up to 1,000 yards. But time-fuses were very unreliable, particularly those of the French, and it was the exception rather than the rule for shells to explode at the desired moment. A British invention, named after its inventor Major Henry Shrapnel, was 'spherical case-shot', a metal-ball containing musket-shot capable of exploding in the air and showering those below with the small musket-balls; its fusing was more reliable and it had a longer range than shell. The most effective artillery projectile was grape or canister-shot, a canvas or metal container of musket-balls or scraps of metal that belched forth from the mouth of the gun in the manner of a sawn-off shotgun; devastating in effect, its range was only about 300 yards.

It did not take a man as discerning as the Duke of Wellington to realise that any or all of these projectiles would be more effective against men packed in column than against those extended in line, in the British style. Also, he was aware of the destructive power of the personal arms of his infantry, knowing that 500 men in the two-deep line favoured in the Peninsular War could fire between six and ten shots per yard per minute, to decimate the small, closely-packed target represented by the head of a column which, even if formed of a French division of 5,000 men advancing on a two-company front, could only bring 340 muskets to bear. The very foundations of warfare of the period rested upon the basic factor of the man-in-the-ranks and his musket which, in the hands of the British soldier backed by high degrees of discipline, training and courage, was a highly effective weapon honed by practice and use.

The British smoothbore musket, known as the 'Brown Bess', fired a standard No. 11 bore 0.76-in. calibre ball 13½ to 14 to the pound; trained men could fire the musket four times a minute, but it was inaccurate at ranges over 80 yards. It took a bayonet of about 15 inches in length which, when fixed to the muzzle, greatly interfered with rapid loading. The other main British weapon was the Baker rifle, issued to the British Rifle Corps; it fired a ball 0.615 calibre and could only be fired twice a minute as its rifling made loading a slower business. However, the rifling also made it immeasurably more accurate than the smoothbore musket of the infantryman, particularly in the cool hands of such experienced marksmen as those of the 95th Rifles and their kind.

The standard French musket in the Peninsular War was the Model 1777 Charleville flintlock musket; calibre .69, twenty-two balls to the pound, giving a slightly longer range than the Brown Bess, but with less impact. French musket-balls fitted British weapons, and were sometimes used in emergencies (Wellington noted such an occasion in 1813) but captured British ammunition could not be used in French muskets.

This last fact might well be the subject of a 'Chance Card' or a 'Military Possibility' in a wargame controlled by rules that include the supply of ammunition – both sides run-out, but the British prevail by using balls taken from the pouches of dead Frenchmen!

8 THE FIGHTING ASSESSMENT OF THE ARMIES OF THE PENINSULAR WAR

Every army throughout military history has been controlled and affected to varying degrees by those inter-related factors which have governed the fighting ability of soldiers since the beginnings of time. Co-ordinated, the sum total of these qualities made an army the good, bad or indifferent force revealed by history. For wargaming purposes in this book, these qualities are evaluated on a chart, each being given a points value that is totalled to provide a Fighting Assessment figure for each army. By allocating a points value of Above Average – 3 points; Average – 2 points and Below Average – 1 point to indicate specific superiority of inferiority in essential factors, Fighting Assessment Charts are intended as a guide to the formulation or adaptation of rules controlling warfare of the period.

A Peninsular War wargame. Figures and terrain by Peter Gilder.

As an example, take a typical Fighting Assessment Chart – that of the Peninsular War of 1808–1814. The British overall commanders Wellington, Graham and Beresford were all first-class men while their French counterparts – Joseph, Jourdan, Junot and Massena must be rated second-class. The British subordinate commanders – Pakenham, Hill, Picton, Cole, Alten and Hope were also first-class while history indicates that the French subordinates – Soult, Marmont, Ney, Bonnet, Clausel, Suchet, Gazan, D'Erlon, Reille, Foy and Murat should be rated second-class. The two armies are both considered to be first-class in composition, with French efficiency balanced by the superb British Light Divi-

THE PENINSULAR WAR 1808–1814

Factors	BRITISH		FRENCH	
	Pts.	*Remarks*	*Remarks*	*Pts.*
(a) Commander	3	Wellington, Graham, Beresford	Joseph, Jourdan, Junot, Masséna	2
(b) Subordinate Commanders	3	Pakenham, Hill, Picton, Cole, Alten, Hope	Soult, Marmont, Ney, Bonnet, Clausel, Suchet, Gazan, D'Erlon, Reille, Foy, Murat	2
Composition of army	3	Including Portuguese, King's German Legion and Morillo's Spanish		3
Mercenaries				
Training	3	Light Division superb		3
Discipline	3	Cavalry weak		3
Morale	3			2
Religious fervour				
Tactical system	3			2
Tactical formations	3			2
Speed of manoeuvring	3			2
Arms and equipment	3			3
Weapon handling	3			2
Special Factors		Portuguese and best Spanish troops classified equal to British, *when supported by them*	−1 when operating in areas of guerilla activity	
	33	*Total*	*Total*	26

sion, their weak cavalry and the fact that part of Wellington's force was Portuguese and Spanish. When supported by the British – the Portuguese and the best Spanish troops are classified as equal to their British comrades. Results make it reasonable to give Wellington's men a first-class morale rating with the invariably defeated French (also beset by guerillas), given a second-class valuation. First-class tactical systems, formations and speed of manoeuvring must be granted to the British, whilst the French, so frequently out-manoeuvred by Wellington, are second-class. The arms and equipment of both sides are first-class but by comparison the weapon handling of the French is second-class as opposed to the British first-class (probably due to the excellent Light Division). All these varying factors add up to a Fighting Assessment total of *33* for Wellington and his men against *26* for their French opponents, whose points total suffers when they are operating in areas of Spanish guerilla activity.

This chart is compiled by comparing values for the principal military attributes of both sides, classified under the following headings:

1. (a) Commander (b) subordinate commanders

A marked feature of many of the campaigns of the period was the degree of equality between opponents, so that victory went to the side with the most talented or competent commander. It must be clearly understood that the grading of commanders and their subordinates does not necessarily indicate that the competence and morale of the man in the ranks (including junior officers) was the same as that of his commanders, although his performance was greatly affected by the morale and ability of those leaders.

2. (a) Composition of army

This is an attempt to represent the 'balance' of a force – whether it has reasonable proportions of infantry, cavalry and artillery, plus whether or not that force is adequate for the role it is called upon to play. For example, Wellington's only reverse, at the siege of Burgos, was mainly because of his lack of sufficient heavy siege artillery.

(b) Mercenaries There was no significant or noteworthy employment of mercenary troops during the period covered by this book.

(c) Training There are exceptional occasions, such as at Baylen on 19 July 1808, early in the War, when an ill-trained army of Spanish patriots, swept along by revolutionary fervour, overcame trained and disciplined opponents. But these isolated instances should not obscure that fact that the best trained armies are most likely to win, because their soldiers react better and more rapidly to orders; they have a greater resistance to stress and, because training has broken them of the habit of questioning their superiors – they obey orders without question.

(d) Discipline A badly trained army is usually a badly disciplined force. All other things being equal, a force with a strong and respected commander and men who have confidence and disciplined control over themselves is likely to perform better than a rabble who wander off in search of loot or food, and soldiers who go home when it is harvest time. After the Battle of the Nivelle, Spanish misbehaviour forced Wellington to send back over the Pyrenees all Spanish forces except the better disciplined battalions of Morillo, who took part in the subsequent operations in France, around Toulouse.

3. (a) Morale

Morale is the 'character' of soldiers as expressed in their reaction to stress and strain, and it is the factor which determines the specific point at which an army or a unit is no longer capable of carrying out its duties and withdraws from the battle, either in good order or as a fleeing mob. A series of defeats will gravely lower the morale standards of a force, while the presence on a battlefield of an outstanding leader, such as Napoleon, can severely affect the morale of his opponents. An army carried along on the tide of victory and led by a popular and trusted commander has such high standards of morale that they can withstand reverses or losses that would normally mean defeat. Wellington's army in the Peninsula is an excellent example of this.

(b) Religious fervour Not present during any of the military operations considered in this book.

4. (a) Tactical system

Tactics are the art of disposing military forces on the battlefield and in actual contact with the enemy. This section on the Chart seeks to assess and classify the battlefield tactics employed by the army under review.

(b) Tactical formations Wellington's use of infantry in line covered by riflemen to defeat French column attacks was a first-class tactical formation. On the other hand, during the Napoleonic Wars there were numerous instances of French column attacks being highly successful against opponents less steady than the British infantrymen.

(c) Speed of manoeuvring One reason for the success of the early French Revolutionary Armies was because they marched faster. Throughout the Napoleonic Wars, the French armies marched faster than the formations of most other countries, who were slower to manoeuvre and take up positions, besides frequently being surprised to find the French far closer than they had anticipated. Good leaders working to efficient tactical systems and commanding well trained and disciplined forces invariably manoeuvred in a speedy manner.

5. (a) Arms and equipment

There was surprisingly little technical improvement in both hand weapons and artillery during the period covered by this book, in fact it was only from the rifled hand weapon (the Baker rifle) onwards that the infantryman was armed with a weapon superior in accuracy, range and rate of fire to the longbow used at Crécy in 1346. Even the famous Brown Bess musket did not fire much faster or more accurately than the handguns used by the Spanish some 200–250 years earlier – although at close range it was a deadly weapon, its inaccuracy made it necessary for soldiers to fire in ordered volleys. The mode of employment of artillery varied considerably during the Napoleonic Wars by such methods as Napoleon's massed batteries in his later campaigns. But the cannon itself did not really improve greatly until the last part of the 19th century, when grooved, breech-loading guns were introduced.

With the weapons remaining more or less constant, those occasions on which one army had a marked advantage in this aspect over their opponents are recorded and should be reflected on the wargames table. Such an example is the battalions of the 95th Rifles, armed with the Baker rifle; the other units of the Light Division carried the customary musket.

(b) Weapon handling With the smoothbore musket, all infantry were theoretically equal but some (such as the Light Divison) were more equal than others! Rules must be so formulated as to make this fact evident on tabletop battlefields.

6. Special factors

These should only be included where they had a material effect

42nd The Royal Highland Regiment (The Black Watch) in a wargaming reconstruction of the Battle of Busaco 1810.

Wargaming the Peninsular War. British riflemen and French Tirailleurs engaged in a firefight in mountainous country. (Featherstone collection)

upon the course of the operations. During the Peninsular War the French army were plagued by guerillas, harassing their flanks and cutting their lines of communication, which should lower their points score.

Because, from a realism and authenticity point-of-view, these factors are vital when matching one army against another in a tabletop campaign, they must be carefully considered and reflected in rules and conditions controlling a battle or a campaign.

The Fighting Assesement Chart should reveal sufficient reason for one army to be victorious and the other to be beaten. Thus, when wargaming, if rules are accurate and produce authenticity, then the side that was defeated in real life should again lose, unless its ranks are weighted by giving it the additional strength indicated by disparities in the respective Fighting Assessment Figures. In conjunction with the grading of a commander into 'Above Average', 'Average' or 'Below Average', a Fighting Assessment Figure enables the wargamer to obtain a balanced idea of two forces if they are to be reasonably matched. For example, an army with a Fighting Assessment Figure of, say, 35 would overwhelm an army numerically equal but with a Fighting Assessment Figure of 25. With the aid of Fighting Assessment Figures, it is possible to work out the respective numerical strengths required in two armies for them to be sufficiently well matched so as to give a balanced and enjoyable wargame.

The numerical superiority allowed a lower rated force need not necessarily consist of increased numbers of figures. It can be worked out in points representing each battalion of infantry, squadron of cavalry or piece of artillery at the rate of infantry regiments at 50 points each, squadrons of cavalry at 45 points and guns at 15 points. This means that an army, although lowly rated on a Fighting Assessment Chart, may be known to be strong in artillery so that their guns are increased to bring the weaker force up to some degree of parity with their stronger opponents. Thus, the points difference may be made up (a) to an historically realistic choice or (b) to the wargamer's personal preference, taking terrain and circumstances into consideration.

Compare the total figure-valuation of the two Peninsular War armies – where a British/Portuguese army with a Fighting Assessment Figure of 33 is opposing a French army with a Figure of 26, so that the British are about a fifth superior to the French who, to present a balanced opposition at parity, need strengthening numerically by a fifth. This 21% difference means that a balanced French army should be a fifth stronger than Wellington's army, requiring say 500 men to take on a British/Portuguese army of 400. That is, of course, assuming that the rules are so formulated as to actively demonstrate those British advantages and French failings that gave rise to the original Fighting Assessment Figures of 33:26.

Another method of balancing forces is to allow a superiority shown by the Fighting Assessment Figure to affect the power of a formation. Hence, to all firing or morale factors add ONE for every 10 parts of the Fighting Assessment Figure, beginning at 11. Thus, an army with a Fighting Assessment Figure of 35 would have 1 point for that part of the Figure between 11 and 21; a second point from 21–31 and a third point for the 4 remaining of the Figure. This gives a total of 3 points to be added whenever weapon, firing, tactical or morale factors are being considered.

However, there is little value in having different values for armies if the rules under which they are fighting are not so formulated as to reflect these varying powers. If a Fighting Assessment Chart reveals, as it does for the Peninsular War 1808–1814, that the British commanders were superior by a rating of 3:2 to their French counterparts, then the rules for the campaign must be slanted to reflect that fact by allowing the British to transmit their orders and act upon them far more quickly than the French. If we know that the French fought in column and the British in line, then the rules must make them do so on the wargames table.

To provide a balanced battle, the wargamer may consider certain other factors. For example, when the defending army is in a strong position, behind fortifications or earthworks, how much stronger should be the attacking army?

If a Fighting Assessment Chart indicates that a force is armed in a vastly superior manner to its opponents (as were the French when fighting the Spanish armies) then rules must be so devised as to increase the effects of these superior weapons. The well-known Wargames Research Group and other lesser-known sets employ weapon and tactical factors that indicate the superiority (or otherwise) of the involved forces, under designated conditions. Weapon and fighting factors provide a series of plus or minus figures that reflect the power of the weapon concerned; the opponent and his weapon; plus such tactical factors as whether a unit is distracted or disorganised, whether it has received casualties itself or is firing upon a moving unit etc. The resulting total, realistically affected by a dice throw to reflect the fluctuations of war, when applied to a Casualty Calculation Chart, reveals the number of enemy casualties. If, before consultation of the Casualty Calculation Chart, the numerical representation of weapon and tactical factors has added to it a further figure reflecting the Fighting Assessment value of the unit concerned, then the resulting casualties should realistically reflect the power of that unit. A force noted for its superior handling of weapons (such as the 95th Rifles in the Peninsula) can be benefited by the same adding/subtracting method.

Fighting Assessment Figures can have a bearing upon orders given to armies in that the force with a high Fighting Assessment Figure is allowed to write orders each move and use an alternative order system; a force with an average Fighting Assessment Figure is allowed to write the orders every third move and may use *some* alternative orders. Armies with a low Fighting Assessment Figure must write initial orders at the start of the battle for the entire game.

Because it is very difficult to do in any other way, these Fighting Assessment Charts are worked out on the basis of a 'neutral terrain' but they can be amended when a force was operating on terrain that it found more suitable to its style of fighting, with perhaps the advantage of a friendly local population. The 'Peninsula' chart makes allowances for the French being affected by guerilla activities. The terrain over which a battle was fought could be decisive, particularly where one army was fighting under conditions peculiarly suitable to itself, a classic example being the Russians during the French retreat from Moscow in 1812.

The Fighting Assessment Chart given here does not include a Battle Effectiveness Figure denoting the stage at which a unit (or an army) is so demoralised by lowering morale and casualties that it is unable to carry on fighting and withdraws from the battle. Military history provides countless examples of such occasions, the principal reason for battles ending before one side or the other is completely wiped out. All units can be graded before the wargame begins, with Guard or élite troops being given a lower Battle Effectiveness Figure (i.e. they can take more casualties) than Line, Militia or irregular units. By this method, 'unequal' wargaming can be enjoyed where one army has a greater number of low Battle Effectiveness Figures balanced by irregular and militia units with high figures, opposed to a force made up entirely of Line units with average Battle Effectiveness Figures.

Fighting Assessment Charts consider an army as a whole. In association with Battle Effectiveness Figures, or in lieu of them, the following system provides an indication of the quality of specific units. Make out a card for each unit and on it will be recorded (and amended when necessary) five separate points values, each representing a factor likely to have a bearing upon the state of that regiment. The classifications are:

(A) Type of unit

Points		
5	Corps élite	– Guards regiments.
4	Line regiments	– veterans, experienced troops.
3	Line regiments	– losses made up with young, inexperienced troops.
2	Conscripts.	
1	Irregulars.	

(B) Leadership

5	Exceptional commander, usually victorious.
4	Average, popular commander, experienced.
3	Average commander, lacking experience.
2	Below-average commander, with experience.
1	Below-average commander, lacking experience.

(C) Supply

5	Well-shod and clothed, regularly well fed.
4	Well fed, other supplies not so plentiful.
3	Reasonably clothed and supplied.
2	Irregularly supplied.
1	Prolonged shortage of all supplies.

(D) Physical condition

5	In excellent health, rested.
4	Reasonable health, rested.
3	Reasonable health, need resting.
2	Indifferent state of health and fatigue.
1	Sick, battle-weary.

(E) Mental condition

5	Confident, buoyed by intense patriotism.
4	Reasonably patriotic and optimistic.
3	Reasonably patriotic but pessimistic.
2	Dubious of their chances.
1	Realise their position is hopeless.

Now total up the five figures that indicate their rating in the various classes; divide by 5 to give a total figure that, in comparison with the total figures of other units, will indicate their status.

9 HOW TO AUTHENTICALLY WARGAME AN HISTORICAL BATTLE

It is a most attractive proposition to refight the famous battles of history, to reproduce Waterloo or Gettysburg for example, on the wargames table. Unfortunately, it is very difficult to put this into practice with any degree of realism because such battles – often significant milestones in history – involved many thousands of soldiers fighting over vast areas of ground. In 1815 on the Belgian field of Waterloo, there were 100,000 French, 68,000 British, Dutch, Belgians and Germans, and more than 100,000 Prussians – even scaled down 100 men on the field equalling one man on the wargames table only provides ludicrous 7-man battalions, and the wargamer will require about 3,000 model soldiers. For a major battle, it was fought on a relatively small field, yet it stretched across a frontage of about 4 miles and, even to a drastically reduced ground-scale, requires an enormous table. This inevitable scaling-down of numbers to such a laughable level almost certainly destroys a sense of realism or an accurate result.

The wargamer attempting to simulate the actions of the Peninsular War is fortunate because almost every battle lends itself to practical and realistic reconstruction. This is important because if the wargame is to represent the battle in anything more than its name, the tabletop armies must represent a reasonable proportion of the original formations, both in numbers and types. The Portuguese/Spanish events of 1808–1813 allow this to happen so that, in the case of the two major battles of the War – Salamanca and Vittoria – both can be most realistically reproduced in a manner that conforms to this proviso by using very small-scale figures such as 1:300 Ros-Heroics or the 5mm 'Regimental Blocks' once produced by Miniature Figurines. Backed by such essentials as a chronological programme of events on the historical day, rules that are so formulated as to accurately reflect period features of the weapons and their effects, plus national or local pecularities of the troops involved, then what results should bear a recognisable resemblance. To some extent this is handicapped by the wargamer commanding the side victorious on the day having an advantage through being aware of how and why the historical commander triumphed, although the loser through hindsight knows what NOT to do on this occasion!

To counter this, wargamers must consider what MIGHT have occurred in relation to what DID occur, considering each successive aspect of the battle in its correct context and in true chronological order so that it can be simulated and affected by the fluctuations and fortunes of war, without radically departing from the military possibilities of the day and age. It is doubtful whether wargames will ever give a profound military insight, but the wargamer can hope to achieve an understanding of the problems of the commanders in the field and a glimpse of the military thinking of the time by refighting a battle in the correct tactical manner, using scaled-down formations and weapon-reactions of the day.

The intention of this section is to discuss these factors and to suggest practical methods of simulation that will result in an accurate, realistic and – most important – an enjoyable facsimile of the original battle. Because it was one of the major battles of the Peninsular War, considered by the Duke of Wellington to be his greatest and most satisfying triumph, and because it reveals a multitude of fascinating tactical movements and hard fighting in classical Napoleonic Wars style – the principal example selected is the Battle of Salamanca, fought on 22nd July 1812. Not a big affair by the numerical standards of warfare, but conveniently situated as to make possible its simulation on a scaled replica of the ground, using exactly the same number of formations put into the field by Wellington in his Anglo-Portuguese Army and that of the French Commander, Marmont.

This was made possible by using 5mm Regimental blocks, allowing the 90 battalions of British and Portuguese infantry forming the 7 Divisions plus the Independent Portuguese Brigades; the five cavalry brigades of Stapleton Cotton, formed of 14 regiments; the 8 French divisions of 78 battalions of infantry, plus Boyer's Heavy and Curto's Light cavalry brigades. The only digression allowed was in artillery – Marmont's 78 guns being represented by 32; and the British 60 guns by 24 – on the assumption that, even using the mildest wargaming concepts of the effects of artillery, would result in the full numbers of guns on the table exerting a certain over-emphasis which could unbalance the battle both as a wargame and a reconstruction.

Essentially, the terrain had to closely resemble both in scale and appearance the historical area and a map of the field was studied to find relative distances, to be scaled-down when building the tabletop terrain. For example, the Lesser and Greater Arapiles were 600 yards apart; 800 yards west of them was the lower ridge Los Arapiles; south of it some 1,200 yards away was a low ridge about two miles long, wooded, along which Thomiere's column was marching when hit by Pakenham's 3rd Division, who emerged unseen from woodland 500 yards away to the west; extending in a 'L' shape from the north, the French columns were spread over 4 miles of front; the leading column (Thomiere's) strung-out in line of battalion-columns more than a mile long, imperfectly protected by Curto's cavalry. All these dispositions, scaled-down to an 8-feet-by-5-feet table, at the opening stage when the British 3rd Division came into French view, set the scene for the Salamanca reconstruction that followed.

All historical movements were followed, their timing an essential feature of the battle, made possible by a Time Chart that recorded off-table moves (or concealed moves) of different formations in outflanking, or attacking movements that brought troops onto the table at some intermediate stage in the conflict. Not so marked in this game, a vital role of the Time Chart is to keep detached formations in touch with the commander, perhaps unaware of their exact location, by means of gallopers whose progress is recorded on the Time chart. Their non-arrival or delay provides Military Possibilities that can realistically affect the course of an historical battle

No self-respecting wargamer likes to be saddled with an army in a battle they decisively lost, with few tangible alternatives to reverse the result, although the more spirited player may welcome the chance of revealing greater degrees of generalship than those shown by the vanquished commander, by turning defeat into victory. So he must be given some encouraging chance of changing military history, without making a 'dog's dinner' of the reconstruction. This is done by the use of Military Possibilities, controlled and logical alternative courses of action which, had they been taken at the appropriate moment during the battle, might well have caused a complete reversal of its result. Military Possibilities are courses of action followed by opposing wargamers with the agreement of both; they are NOT excuses for engaging in whims and fancies nor for diverting events merely 'to see what happens'. In some cases the course of action indicated by a Military Possibility results in a more reasonable and credible result than actually occurred on the field of conflict. Some might depend upon Luck, represented by the throw of a die or the turn of a Chance Card, which is the simplest means of simulating the ebb and flow of war. It has to be decided whether Military Possibilities are radically to alter the historical course of the battle or be restricted to relatively minor aspects of it, then they may result in a more interesting 'twist' in tactics, or even influence the eventual outcome.

So decisive was the nature and outcome of the Battle of Salamanca that credible Military Possibilities are few and far between; there was little chance of it being other than a British victory once Pakenham's 3rd Division and Le Marchant's Heavy cavalry had routed their respective opponents. For the French to be in with a chance, one or both of these attacks must fail, and/or Clausel's counter-attack must succeed – three possibilities which, if allowed to occur, must take Salamanca into the realms of fantasy! It might be permissible to make Pakenham less successful to make more of an even battle, or Clausel's counter-stroke could be given a chance of success, which might make the battle a draw but an unlikely French victory. Rubbing it in, a Military Possibility could allow the Portuguese to reach and hold the Greater Arapile before the French arrive! Take heart, Military Possibilities abound in many of the battles described in this book! The interest and colour they will bring to the wargames table are proportionate to the experience and ingenuity of the wargamers.

Salamanca is the classic example of victory going to the better commander, which almost certainly occurs when both forces are equal in strength, morale, equipment, position, manoeuvrability etc. – as was the case on 22nd July 1812. In almost every battle described in this book, a strong enemy force was beaten by a numerically weaker army solely because the victor's commander – Wellington – possessed outstanding tactical leadership and powers of command, being capable of inspiring his men to exceptional heights. Thus, just as history dictates that one commander was 'Exceptional' while his opponent was only 'Average' or 'Below Average', so this fact has to be reflected on the wargames table.

The effect upon the battle of a commander's rating must be reflected by his troops, since those of an 'Exceptional' commander will possess a higher standard of morale and better fighting qualities – represented by adding any pre-decided figure to any dice affecting morale or fighting qualities. Conversely, a 'Below Average' commander would necessitate deducting a similar figure from dice scores; the troops of an 'Average' commander remain unaffected. If a system is in use where, at the start of the battle, orders are written for each army or group, the grading of commanders can be reflected by ruling that those armies or groups with 'Below Average' commanders must conform rigidly to their original orders until they are disorganised by a forced reaction (such as a low morale rating causing retreat or rout), 'Average' commanders can write orders to carry through three moves of the game; at the conclusion of the third move, fresh orders may be written if circumstances have not altered the original instructions. 'Exceptional' commanders may write orders at the beginning of each move of the game. In passing, it is fascinating to note that military history reveals many 'Below Average' commanders to be men of great personal courage, such as General Sir Hugh Gough of Sikh War fame, or Lord Methuen and General Redvers Buller in the Second Boer War. Perhaps bravery was bestowed upon them to compensate for deficiencies in other qualities.

Akin to Military Possibilities but far more mundane in nature, CHANCE CARDS introduce pleasant or unpleasant factors that materially affect aspects of a battle, perhaps even its result. They pose eventualities – tactical, physiological, or psychological – and the commander drawing such a card has to take practical steps to carry out the instructions it bears. Each historical battle can have its own set of cards designed to cover eventualities likely in the context of that battle; every real-life battle abounds with situations which, in a wargame, could give rise to the use of Chance Cards forming the 'human' element that may affect the purely tactical aspects of a Military Possibility. A Chance Card could declare that the horse of the galloper sent by Wellington to order Pakenham to strike Thomiere's isolated division, has gone lame and the order is delayed – with possible consequences that could affect the whole battle. Or, when Le Marchant's Heavy cavalry are charging forward on the right of the Allied 5th Division, they encounter a gully that delays them, or even halts the charge. Remembering Wellington's cavalry at Talavera, the card might decree that Le Marchant does not have his force under the rigid control that history indicates; or the columns of Sarrut's 4th French Infantry Division – the last to be encountered and destroyed by the Heavy Cavalry – have time to do what they could not do in 1812, form square. Chance Cards are those major or minor everyday eventualities that can madden or delight, depending upon who is at the sticky end – but they certainly liven up, even revolutionise, a wargame!

The Battle of Salamanca began at 4.45 in the afternoon, carried on throughout the gathering dusk and was finally fought-out in the dark, ending at 9.30pm. So perhaps the French commander's greatest chance of victory is to slow-up the game to give himself time to extricate as much as possible under cover of night, living to fight again another day. More facts and figures on Salamanca are given in the section of this book dealing with each distinct battle – study this section and those facts, and enjoy what will undoubtedly be a memorable game.

10 REALISTIC RULES FOR WARGAMING THE PENINSULAR WAR

Wargamers are strong-minded people with very definite ideas of the abilities and potentialities of the armies they command. They derive their views and opinions from a variety of sources, but principally from reading, films and television. This is fine in so far as it goes, providing the reader does not blindly accept one source but takes the trouble to read a variety of books on the same aspect and then gauge them against each other before arriving at his own conclusions. It is at this point that the snags begin to occur because, having no first-hand knowledge or personal experience of Horse-and-Musket warfare, the wargamer's conceptions and interpretations of his reading are coloured and affected by his temperament and character.

Then, the tender question of national pride must be considered because the average British wargamer, particularly in the Napoleonic period, tends to slightly overrate Wellington's men. In so doing, the rules he formulates are inevitably tinged with a slight (and perhaps unconscious) bias in their favour. But it is no denigration of our worthy ancestors to accept that Napoleon's French infantrymen were often very experienced and very good and that, in their heyday, the French cavalry were superb. And yet, a realistic set of rules for the French and Austrian campaign in Italy during the last years of the 18th century MUST reflect the fact that the Austrian cavalry at that time were superior to the French, being the best mounted arm in Europe. Similarly, it is written in a number of authoritative places that the Russian cavalry at the time of Borodino in 1812 were as good as any others in existence.

Perhaps it is not unreasonable to claim that the British infantryman *in line* had a better chance of turning back cavalry than did his Austrian, Prussian or Russian counterpart, who had been discouraged by a long series of defeats and disastrous happenings on fields and in battles far bigger than experienced by the British infantryman, until Waterloo. A three-deep line of high morale infantrymen had a *slight* chance of turning back a cavalry attack but no set of rules should make this a right because it was an unlikely and infrequent happening.

Broadly speaking, so far as the Napoleonic period is concerned, the main tactical concept was for infantry to be forced into square formation because of the threatening presence of enemy cavalry; then their tightly-packed formations were flayed with close-range artillery fire from batteries that were themselves out of the range of the infantry muskets. Eventually the infantry ranks were so disordered by this artillery fire that they became vulnerable to the attacks of cavalry, or, even more likely to succeed, infantry in column. Therefore, authentic rules for this period must be so designed that they give cavalry little chance of success against infantry in square, that foot soldiers in the open and not in square have little or no chance of remaining unbroken in the face of a cavalry attack.

Similarly, when recreating the Peninsular War, British infantry should be given some benefit for their known solidarity in line formation but French infantry, not only in the Peninsular War but at all times, must be given benefits when authentically attacking in column. If any battle is to be realistically reconstructed with both armies employed in the authentic tactics of the period, then rules must be so devised as to *force* those tactics to be used on the tabletop, even though the wargamer/commander may be conversant with differing tactics of another period that are likely to be more successful. The information on armies and their leaders, their equipment and their morale state, together with the Fighting Assessment Charts, are sufficient to enable the wargamer to compile rules that reasonably reflect the strength and weaknesses of the armies of this period.

There are certain essentials which must be considered when formulating any set of rules to control a wargame, these include the reflection of an army's strength and weaknesses which, taken in comparison with the opposing army, produce a reasonable simulation of the events of the time.

The rules must set out definitions that accurately reflect the speed and accuracy of an army's movement and manoeuvring, together with a means of simulating their ability or otherwise to maintain the rate of small arms and artillery fire that is known to be historically correct. For instance, under favourable conditions, an artillery battery during the Napoleonic period took a minute to come into action and from one to three minutes to limber and move off from a firing position whilst a single gun could unlimber and fire in thirty seconds. A trained British soldier took between twelve and fifteen seconds to reload and fire his Brown Bess musket; French authorities believed that two shots per minute could be fired from the musket of the period. The musket itself was inaccurate over ranges from more than about 60 yards and there are recorded instances of volleys fired by entire battalions only hitting four men at ranges of 100 yards. It took an army of 30,000 men two hours to form line of battle if not under fire but longer if some of them were already engaged; infantry charges took 60 seconds to cover 180 yards while cavalry was said to be able to charge for 700 yards during the same time – in either case the infantry being attacked would be lucky if, under stress, they could fire more than one round before the enemy was upon them.

It is essential to bear in mind that these movement speeds and rates of fire were recorded during practice or drill when no one, including both the men and the recorder, was under any stress. The times are for trained veterans whose morale was steady; nervous troops, conscripts or inexperienced soldiers were inferior in every respect and took much longer to move and fire. Herein lies one of the great pitfalls for the inexperienced wargamer in that he accepts these claims and figures to formulate rules that allow unjust powers to favoured forces; it must be realised (and rules must so reflect) that under the stress and strain of battle soldiers did not perform so creditably.

The wargamer should never overlook that he is attempting authentically to recreate a period battle on a grossly under-scaled battlefield, using similarly under-scale numbers of inanimate miniature figures. He manoeuvres his armies in a style and manner much influenced by his temperament, military knowledge and any peculiar concept of historical tactics he may have derived from research. His little warriors cannot march and fight of their own volition so they have to be tactically directed by their wargamer/commander. But he possesses military knowledge of a much later date than the period in which he is specialising, so that the tactics employed by his armies are those of leaders in far more advanced periods of military history, well beyond the military knowledge of their historical commanders. Consequently, although researching and painting armies to represent a reasonable scaled-down simulation of specific armies, those forces are performing tactical manoeuvres and evolutions arising from a military hindsight denied to their commander in those far-off days. A well-conceived set of rules, by formulating conditions and rulings that reflect the particular attributes of specific armies, can do much to correct this anomaly.

It is at this point that one encounters perhaps the greatest obstacle to wargaming becoming a majority-hobby. Wargamers, usually lacking professional military training or actual battlefield experience, acquire from reading and research widely differing concepts of tactics and warfare that are almost irreconcilable. They laboriously formulate rules that emphasise their accurate or possibly misguided concepts while viewing with ill-conceived disgust and frustration other people's rules that do not agree with their own concepts. In such company, wargamers have been known to be reluctant to perform some rather brilliant and quite authentic tactical manoeuvre simply because of an awareness that fierce arguments will be required to support its legality!

The greatest error into which a wargamer can fall is to select an army and then tactically manoeuvre it in a manner that bears little relation to its historically recorded tactics. For example, when refighting the Peninsular War of 1808–14, it must be accepted that one of the predominant tactical reasons for the Duke of Wellington's success lay in his chosen method of forming the well-trained and confident infantry that comprised his army into double-lines on the back slopes of a ridge. This tactic enabled them either to advance to the crest or wait until the advancing French columns came into view, before pouring a superior volume of musketry into their packed ranks prior to charging the discomforted column. And, equally important, his valuable infantry, behind the ridge-crest, were sheltered from enemy artillery fire. This was something of a precedent in an age when few commanders – not even Napoleon – drew up their formation so that they were out of enemy artillery fire.

The majority of tactics have historical precedents and those described above resemble the style of fighting of the Hundred Years War (1346–1453), when consistently the French were devastated by advancing on foot into English archery volleys. To fight the Peninsular War in any other way is to prostitute history, and to create a situation completely lacking in authenticity.

The style-of-fighting of the armies of the Peninsular War period are detailed in these pages, and their commanders are classified from Bad to Brilliant. Unequal-strength forces and the grading of commanders are balancing factors that persuade a wargamer/commander to accept that what he gains on the swings he loses on the roundabouts. Few wargamers will feel inclined to manoeuvre their armies in a tactical style that is historically correct when hindsight tells them that those tactics invariably led to defeat. The French wargamer/commander is going to be very disinclined to throw his columns up the slopes of a ridge when he is aware that the English/Portuguese line, with all its devastating firepower and subsequent bayonet charge, awaits him below its crest. Herein lies a certain irony because this seems to be a regular historical occurrence!

To bear any realistic similarity, other than title, to the actual historical conflict, tabletop forces MUST be made to fight in a tactically accurate manner, dictated by the specific battle's historical background.

All wars, whether they be real-life conflicts or miniature affairs on tabletop battlefields, are fought to rules, and those used to control the tactical manoeuvring of model soldiers, far from being standardised as for chess, usually reflect the character, temperament and knowledge of their deviser. There are almost as many sets of rules as there are wargamers, all optimistically formulated to give the game the greatest possible realism, which often results in bulky folders requiring reference taking up more time than the actual game! This is quite unnecessary and an equally realistic game – certainly a faster and more enjoyable battle – can be played by concentrating on the key word – MORALE. Accepted from a military point of view to be the psychological condition of soldiers as it concerns confidence and discipline, success in battle being achieved by the winners possessing higher standards of both than the losers. However, regardless of the standards of morale at the start of a battle, it can be materially reduced through casualties and the reactions caused by combat. This positively implies that the principal object of fighting an enemy, both by missile-fire and by hand-to-hand fighting (mêlées) is to reduce his morale until it is lower than your own, causing him to react in a manner that is detrimental to him.

This inevitably leads to the proposition that, when wargaming, it is expeditious to formulate rules so that they achieve their ends by being balanced on the State of Morale of the respective armies. There is little mileage in calculating numerically the precise effects of taking fire, whether it be artillery or small-arms, or being charged and mêlée'd – all that is required is an awareness of a unit suffering a lowering in their morale through these eventualities. Or conversely, whether their morale has risen through elation at having lived through fire or a charge – in both cases the rules should reflect these situations by enforcing a reaction, adverse or beneficial respectively, upon the affected unit. Eventually, in this way, one side wins and the other loses.

It is simply carried out by merely combining the effects of receiving missile-fire or taking a charge – giving both the overall title of COMBAT, with the chances of success of a specific type of COMBAT being represented by odds, as a percentage which has to be exceeded when throwing dice if the affected unit is to avoid an adverse morale-reaction. Perhaps the classic example is one peculiarly suited to the Napoleonic period and warfare in the Peninsula, where infantry were easy prey to cavalry when in loose or open order; a little safer when in line; but almost completely safe when in square formation. Thus, infantry caught in the open, in loose or open order need to throw a pair of percentage-dice (20-sided dice, black are 100s and red are 10s, capable of producing a combined score of from 1 to 100; the two 0's together representing the latter) and score 80 percent or more to be safe. When in line they need to score 75 percent; but when in the comforting formation of a square they only require 15 percent – while the opposing cavalry have to throw 85 or over to break the square.

Each army or force is listed on a chart in columnar form, extending from each unit is a series of squares, five for ordinary run-of-the-mill units and perhaps six or seven for élite, veteran or Guard units. When a unit fails to make its required score when throwing the dice, it is considered to be momentarily shaken, so one attrition-point (represented by a tick in the first square) is marked against it; subsequently, it has to throw again – this time a single dice – which, if falling 1 or 2 means that its morale is rock-bottom, it has been routed in disorder and – next game-move – it is pulled back its respective move-distance and stands in disorder until such time as its owner – throwing dice at the beginning of each move – scores 50 or over, when it is considered to have rallied. Each failure to score 50 means a further attrition-point, and the initial dice score of 1 or 2 has already marked-in two more ticks. If, when throwing the single die after an initial 'shaking', scores of 3 or 4 are secured, then one more attrition-point goes in the chart and the unit – next move – goes back its move-distance in good order; it may come forward again on the following move. When all five squares are filled with attrition-points, then the unit is removed from the wargames table. The chart given here as an example is that of Craufurd's famed Light Division, Wellington's 'crack' infantry force in the Peninsula; as they are an élite formation the chart is marked into six squares, giving them one more 'life' than an ordinary line-regiment.

This is the basic idea, the foundation upon which this rules system works, and it can be modified to suit ANY period, any army, or type of troops. This means that a club or group of wargamers can move rapidly from one period or type of warfare to another, knowing that the basic rules are the same with percentages and attrition-points devised to simulate the specific weapons and fighting-styles of the period in use. The system bears the merit of not having to assess individual casualties nor to remove them singly from the table, leaving a unit with perhaps two or three men; *all* reactions and results concerns units AS A WHOLE. This is a great comfort when thoughtfully considered, because all those precious little model soldiers, so painstakingly painted, do not spend much of their time in casualty boxes under the table, but can be permanently fixed in groups (say threes) on cards. Because units are never broken up during a game, they can be of any manageable size – from ten to fifty, they fight as a unit irrespective of their numbers.

Recalling what is sometimes forgotten in the stress of wargaming – that it is a GAME played for amusement and relaxation – the wargamer can be happy in the knowledge that these rules banish calculations and the need to remember anything – one just gazes at the chart on the wall! This is what will be seen:

MOVEMENT:

	Infantry	*Light Infantry*
In column	– 7 inches	When deployed in open order
In line	– 6 inches	– 9 inches

Cavalry

Light	In column	– 9 inches	*Heavy*	In column	– 8 inches
	In line	– 12 inches		In line	– 10 inches
	Field Gun	– 6 inches		Horse Gun	– 12 inches

Changing formation – takes up half a move.
Form or unform infantry square – takes a full move.
Unit formations –
a. Both infantry and cavalry advance in COLUMN forming line on confronting enemy.
b. British infantry fight in LINE
All cavalry fight in LINE
French infantry fight in COLUMN but may firefight LINE
Spanish infantry fight only in COLUMN
c. *Movement & Firing*
Field-guns cannot move AND fire in the same move.

Infantry can half-move AND fire muskets. Light infantry cannot move AND fire Baker rifles in the same move.

Horse-guns take – 2 inches to limber; 2 inches to unlimber; 2 inches to load. Remaining distance may be moved.

d. Movement in woods – Light infantry move ½ rate
Impassable to artillery
Infantry/Cavalry ½ move and are in disorder.

Disordered troops cannot initiate aggressive action; may defend themselves if attacked by seeking to make their required score BEFORE the enemy, who only need to try if disordered troops are successful. If fired upon when in disorder, target-unit takes a mandatory attrition-point (1 tick) for being disordered, then throw single die to check their reaction i.e. score 1 or 2 they are routed (2 ticks); score 3 or 4 go back move-distance in good order (1 tick).

RANGES

	Short Range	*Long Range*
Musket	4 inches	9 inches
Baker Rifle	6 inches	12 inches
Field Gun	10 inches	40 inches
Horse Gun	10 inches	30 inches

UNIT REACTION TO EFFECTS OF COMBAT (MISSILE FIRE OR MÊLÉE)

1. *Failing to make the required score*:

a. The unit is SHAKEN and marks 1 attrition-point on its chart.

b. Throws a *single* % die.

c. Score of 1 or 2 means the unit has been routed; they move back (on the following Game-Move) their move-distance IN DISORDER. Mark 2 more attrition-marks on their chart. While dis-

EFFECTS OF ARTILLERY AND SMALL-ARMS FIRE

TARGET UNIT	BEING FIRED ON BY						
	INFANTRY						GUNS
	In Line	In Column	In Open Order	In Square	In Hard Cover	In Soft Cover	
Infantry Line	30 / 45	20 / 30	20 / 30	20 / 30	35 / 50	30 / 45	40 / 60
Infantry Column	35 / 50	25 / 35	25 / 35	25 / 35	40 / 60	35 / 50	45 / 70
Infantry Open Order	25 / 35	15 / 25	15 / 25	15 / 25	30 / 45	25 / 35	35 / 55
Square	35 / 50	25 / 35	25 / 35	25 / 35	40 / 60	35 / 50	50 / 75
Hard Cover	20 / 30	10 / 15	10 / 15	10 / 15	15 / 25	20 / 30	30 / 50
In Soft Cover	25 / 35	15 / 25	15 / 25	15 / 25	30 / 45	25 / 35	35 / 55
Cavalry	30 / 45	20 / 30	25 / 35	35 / 50	40 / 60	35 / 50	45 / 70
Gunners	25 / 35	20 / 30	20 / 30	20 / 30	35 / 50	30 / 45	45 / 70

First (lower) figure is score required at LONG range
Second (higher) figures is score required at SHORT range
Guard or Elite units DEDUCT 5 from all the above scores.

Target unit must throw % die and EQUAL or BEAT these scores for them to be O.K. Failing, marked 1 Attrition-point; throw single % die – score 1 or 2 unit is ROUTED: score 3 or 4 unit moves back their move-distance in good order and may resume next move.

EFFECTS OF MÊLÉES (HAND-TO-HAND COMBAT)

The unit in the first column on the left, when opposed by one of the formations heading the columns, need to throw a pair of % Dice and equal or beat the number shown in the chart. Thus, *infantry in square* facing *Heavy Cavalry* (Col. 8) need to score 15 or over to be safe. Guard/Elite units deduct 5.

	Line	Column	Open Order	Square	Hard Cover	Soft Cover	Light Cavalry	Heavy Cavalry	Gun Crew
Infantry Line	50	60	30	40	70	60	65	75	30
Infantry Column	40	50	30	35	70	65	65	75	25
Infantry Open Order	70	75	50	50	75	50	75	80	20
Infantry Square	60	65	50	—	—	—	20	15	—
Infantry Hard Cover	30	30	25	—	—	—	20	20	—
Infantry Soft Cover	40	35	50	—	—	—	30	30	—
Light Cavalry	30	35	25	80	80	70	50	60	20
Heavy Cavalry	25	30	25	85	80	70	40	50	20
Gun Crew	70	70	60	—	—	—	80	80	—

Failing to equal or beat the stipulated score, the unit marks on its chart 1 attrition-point for being SHAKEN; then throws a single % die; scoring 1 or 2 means that unit is routed back its move-distance in DISORDER (2 more attrition-points); score 3 or 4 unit goes back its move distance in good order and may resume moving next Game Move (1 more attrition-point for going back).

ordered, they may be fired upon or attacked – then they mark another 1 attrition-point, and then throw a *single* die, trying to avoid scoring 1, 2, 3 or 4 (the penalty scores). If they fail, then they move back in rout or good order, depending upon the score thrown, marking more attrition-points on their chart. To revive from rout, the unit requires to throw 50 on two % dice, which may be thrown at the start of every move subsequent to their rout. Each failure means another attrition-point and a further move back. Scoring over 50, the unit takes that move to rally, and may then resume action on the following move.

d. Score of 3 or 4 means the unit moves back its move-distance, but in good order, on the following move. On the move after that,

SPECIMEN UNIT CHARTS FOR PENINSULAR WAR FORMATIONS

1. Packenham's 3rd Division (at Salamanca)					
	1	2	3	4	5
Alexander Wallace's Brigade 1/45th Regiment of Foot					
74th Regiment of Foot					
1/88th Regiment of Foot					
5/60th Regiment of Foot					
J. Campbell's Brigade 1/5th Regiment of Foot					
2/5th Regiment of Foot					
2/83rd Regiment of Foot					
94th Regiment of Foot					
Power's Portuguese Brigade 1st Bn. 9th Line Inf.					
2nd Bn. 9th Line Inf.					
1st Bn. 21st Line					
2nd Bn. 21st Line					
12th Cacadores					

2. Foy's 1st Division of Marmont's French Army (at Salamanca)					
	1	2	3	4	5
Chemineau's Brigade 1st Bn. 6th Leger					
2nd Bn. 6th Leger					
1st Bn. 69th Line					
2nd Bn. 69th Line					
Desgravier-Berthelot's Brigade 1st Bn. 39th Line					
2nd Bn. 39th Line					
1st Bn. 76th Line					
2nd Bn. 76th Line					

3. The Light Division (commanded by C. Alten, at Salamanca) Elite Formation						
	1	2	3	4	5	6
Barnard's Brigade 1/43rd Regiment of Foot						
2nd 95th Rifles (part)						
3rd 95th Rifles						
1st Cacadores						
Vandeleur's Brigade 1/52nd Regiment of Foot						
1/95th Rifles						
3rd Cacadores						

they may resume action. Withdrawal in good order costs another attrition-point.

e. Every Line unit has FIVE squares on its chart, Guard and Elite units have SIX. An attrition-point (a tick) fills *one* square, so that when a unit has its entire quota of FIVE squares filled with attrition-points, that unit is considered to be SHATTERED and is removed from the field.

At the beginning of the wargame, total the number of units, guns etc. that form an army; each of them will have a potential destruction-point of FIVE attrition-points. Thus, ten units have a collective destruction-point of 50. Assess the acceptable percentage of casualties your army can sustain – an optimistic view might be 50 percent, which amounts to a figure of 25 attrition-points which, when reached, means that your army has taken all the casualties it can handle, and has lost the battle.

The Peninsular War is a very convenient campaign to reconstruct on the wargames table because Wellington's unbroken victorious trend makes it possible to follow the historical sequence of battles in correct chronological order. This is difficult to achieve when wargaming a real-life campaign because if the enemy – on the wargames table – reverses the historical result of a battle, then the whole campaign is thrown out of sequence. This is because it makes it impossible to fight the action historically following on the authentic site or under the known historical conditions, such a course being affected by the 'wargaming losers' being perhaps driven to a distant area, or suffering losses negating the known background to the battle. Thus, instead of an army conveniently progressing to the next historical field as they did in the past, in a tabletop re-enactment, they will be pushed back to another area on the campaign-map, ostensibly to fight a battle that never took place, on an unknown field.

Fortunately, there is a method that obviates such painfully unrealistic courses of events – the Campaign League Table. The campaign is refought in exactly the same sequence as historically occurred, each battle of history following the other, regardless of which side wins on the tabletop fields. The trick is to give the winner 2 points, the loser nothing, with 1 point each for a drawn battle; the points being totalled as the campaign progresses with a perfectly clear and obvious state of affairs at all times. For example, at Rolica it is necessary for the smaller French force in their strong hilltop position, to hold off the considerably stronger British force, and to be able to make their getaway before the arrival of the flanking British detachment under Ferguson. During the wargame, if the French are able to do this, then the British will gain 2 points for the victory, while the French will gain 1 point for holding off the British for a pre-agreed number of game-moves.

11 THE BATTLES AND SIEGES OF THE PENINSULAR WAR 1808–1814

In December 1807, a French army under Junot invaded Portugal. In March 1808, Murat led an army of 100,000 into Spain to place Napoleon's brother Joseph on the throne. On 1st August 1808, Wellesley's army landed at Mondego Bay, north of Lisbon, and within three weeks British forces in the Peninsula received their first 'blooding'.

THE BATTLE OF ROLICA – 17 AUGUST 1808

Opposing forces

Lieut. General Sir Arthur Wellesley commanding 6 infantry Brigades (14½ battalions); 2 squadrons cavalry (240); 18 guns. *Total*: 14,000 (13,500 British infantry (only 5,000 actually engaged) 2,000 Portuguese Auxiliaries (not engaged)).

General Delaborde, commanding a French force (part of Junot's Army of Portugal) consisting of 5 battalions infantry (4,000); 3 squadrons cavalry (300); 5 guns. *Total*: 4,400.

After skirmishing between French tirailleurs and British Light companies, Delaborde retired from his strong position before the village of Obidos in hilly and broken terrain, to a position further south on a horseshoe of hills crowned by a dyke forming a natural breastwork. The French line was about ¾ mile long, flanks resting on deep gullies, fronted by rough ground too steep to climb except at four boulder-strewn gullies. Wishing to defeat Delaborde before a junction with Loison's 9,000 strong force, half a day's march east, Wellesley sent Trant with 1,400 Portuguese through the hills right and west, while Ferguson went left and east with two brigades of infantry, ½ brigade Fane's Riflemen, detachment of 40 cavalry and 6 guns; although part of a pincer to encircle the French, it would protect the flank should Loison

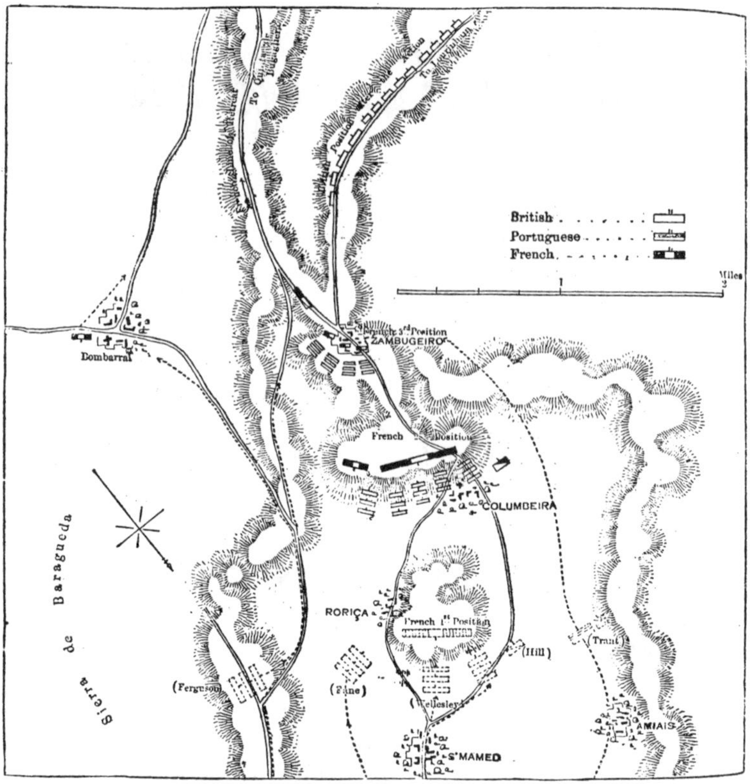

Plan of the Battle of Rolica.

The Battlefield of Rolica. One of the gullies (defiles) up which British troops attacked Delaborde in his second position.

attack. Initially engaging with artillery and skirmishers to give the flanking columns time to advance, Wellesley ordered an immediate general attack after Colonel Lake had been killed and taken heavy losses with a small British column advancing up one of the rocky gullies. Moving upward through the gullies, British infantry were eventually able to form a continuous line along the western half of the ridge, where severe fighting took place. Accepting defeat, Delaborde skilfully withdrew his four infantry battalions, two at a time, covered by cavalry, but lost most of his artillery.

Casualties

Anglo-Portuguese Army – 485; French 700. Although three times larger than Delaborde's force, Wellesley was only able to use about the same number of men as the French in an action proving that the much-vaunted French army were not supermen.

Next day the army marched south to Vimiero, inland of Maceira Bay where a further 4,000 British troops were landed bringing Wellesley's strength up to nearly 17,000. By 20 August they were positioned on two long hills between Vimiero and the sea, where they were attacked by the French the next day.

THE BATTLE OF VIMIERO – 21 AUGUST 1808

Opposing Forces

Lieut. General Sir Arthur Wellesley, commanding 8 infantry brigades (20¼ battalions: 16,312 men); 240 cavalry (2 squadrons); 3 batteries (18 guns). *Total: 16,778*

General Junot (with Generals Solignac, Delaborde, Loison and Margaron) commanding 6 infantry brigades (15 battalions plus Grenadier Reserve of 2 brigades of 2 small battalions each: 10,400 men); 1 cavalry division (1,950); 4 batteries (24 guns).
Total: 13,050

Wellesley had positioned his army in anticipation of an attack from the south, up the Maceira valley; but Junot made his main thrust from the east, against Vimiero Hill, with a right hook against the British left on the Eastern Ridge by Solignac's Brigade later strengthened by Brennier's Brigade. Lacking an overall commander, these brigades attacked separately and ineffectively, while their absence disastrously weakened Junot's main thrust, itself launched piecemeal. To cope with the new direction of attack,

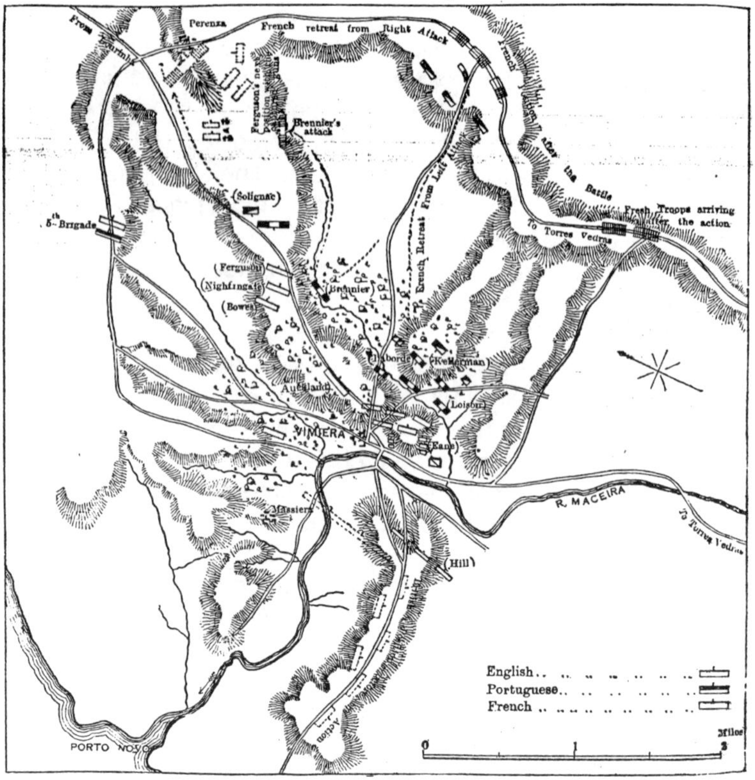

Plan of the Battle of Vimiera.

Wellesley reinforced his centre and left with 4 brigades from the Western Hill, leaving only Hill's Brigade and 2 guns. Vimiero Hill was repeatedly attacked by 8 battalions (13,000) in columns shielded by tirailleurs, supported by mobile field-artillery, with cavalry protecting the flanks. British riflemen so troubled the tirailleurs that they did not adequately protect the columns and the French gunners were persistently harassed so that their fire was light and ineffective. When the moment arrived, the British line, using all their muskets, wrapped their flanks around the columns so that they could not deploy into line under such fire; finally they broke and fled, taking the artillerymen with them and leaving the guns to be captured. This occurred on at least five occasions during the battle, forming a pattern to be repeated in every British success in the Peninsula.

Casualties
Anglo-Portuguese Army – 720 (135 killed; 534 wounded; 51 POWs)
French Army – 2,000 (450 killed; 1,200 wounded; 350 POWs; 13 guns)

This defeat by a relatively inexperienced British force of a veteran French army using tried tactics, led to the notorious Convention of Cintra and Junot's evacuation of Portugal. At this time the political appointment of two inadequate and inexperienced Generals to command the British Army in Portugal caused Wellesley to return to Britain. In July 1808, Sir John Moore had returned to Britain with an army which had been in the Baltic, and was ordered to take them to Portugal. On 6 October 1808, Moore was appointed Commander-in-Chief of the Army in Spain, and on the 27th of that month, led an army of 23,000 into Spain; but at Salamanca in November he was forced to consider retreat to Corunna and re-embarkation. Napoleon, with an army of 194,000, was at the gates of Madrid, the Spanish provincial armies severally defeated, and his mission to aid them made virtually impossible by his own small numbers and Spanish unreliability. An intercepted despatch sent him marching to attack an isolated French force at Saldana on the River Carrion, but learning that Napoleon was hot on his heels, began the epic retreat to Corunna. But not before his cavalry had briskly disposed of some detachments of French cavalry.

THE ACTION AT SAHAGUN – 21 DECEMBER 1808

Opposing forces
Lord Henry Paget commanding the 15th Hussars (approx. 400 cavalrymen) with 2 guns; and belated help from General Slade and 10th Hussars.

General de Brigade Debelle, commanding a brigade consisting of the French 8th Dragoons and 1st Provincial Chasseurs (Hanoverians) estimated at 600–700 men, but more likely 500.

In dreadful weather, with deep snow, slippery roads and heavy frost, Lord Paget, commander of Moore's advance-guard, was attempting a dawn attack to surprise and destroy a detachment of the cavalry screen of Soult's army. Near the road to Carrion de los Condes, it became possible in the darkness and mist of

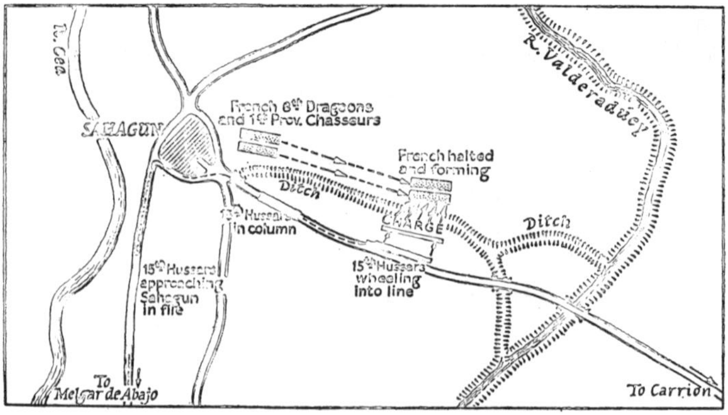

The Combat of Sahagún, 21 December 1808.

early morning to distinguish a cavalry force in close column of squadrons. The 15th Hussars formed open column of divisions at the trot, while the French halted, deploying into close column of regiments, making their formation six deep, then moving parallel they cheered and began firing. The 15th halted, wheeled into line, cheered and advanced on the enemy 400 yards away, crying 'Emsdorff and Victory!' hitting them with such shock that horses and men were overthrown as the 15th cut right through the French column. In a snow-covered area of vineyards intersected with deep ditches, the mêlée lasted about 10 minutes, riding together pell-mell, cutting and slashing at each other. On the bugles sounding 'Rally' the 15th halted and reformed ranks, as the Chasseurs-á-Cheval fled towards Carrion, while the Dragoons retired in the direction of Saldana, both being hotly pursued.

Casualties

15th Hussars had 2 killed, 3 officers and 20 men wounded. An uncertain number of French were killed; 13 officers and 150 POWs and baggage taken.

Sir John Moore skilfully retreated, outwitting and outmarching three French armies under Napoleon, Soult and Junot, fighting rearguard actions in appalling weather conditions in the mountains of Galicia. Eventually, on 11 January 1809 his decimated, exhausted and ragged army of 15,000 reached Corunna with Soult and 20,000 men at their heels.

THE BATTLE OF CORUNNA – 16 JANUARY 1809

Opposing forces

Lieut. General Sir John Moore, commanding 4 divisions of infantry (10 brigades, 30 battalions: 14,800 men); 12 guns.

Marshal Soult, commanding 3 divisions of infantry (8 brigades, 39 battalions: 15,500 men); 3 divisions of cavalry (12 regiments: 4,500 men); 40 guns.

Moore drew up his army on high ground south of the promontory on which the town of Corunna was built, his left flank resting on an impassable river but with an exposed right flank, where Moore posted half his strength in depth. Early on the 16th, Soult attacked strongly against the right, sending feint attacks against left and centre. Covered by concentrated artillery fire, the French attack against the British right-centre resulted in heavy fighting, the village of Elvina changing hands several times, but the French never achieved a permanent penetration. After Sir John Moore was mortally wounded, command devolved on Sir John Hope but before he could exercise any general control, the battle had

Sir John Moore. (After the Engraving by C. Turner)

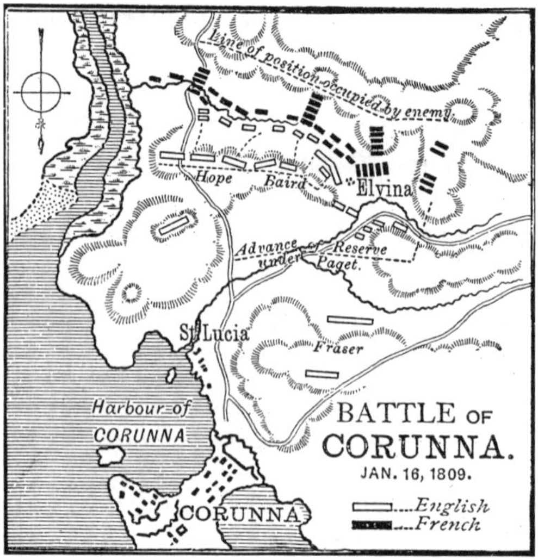

Battle of Corunna.

degenerated into spasmodic long-range artillery fire, eventually drawing to a quiet end with both sides occupying their approximate original positions.

Casualties

British, not definitely known but estimated 800–900 men. The French were said by Fortescue to have had over 600 killed and wounded, although it was generally considered that they lost about as many as the British. The British were able to embark without leaving a man behind. Soult reorganised his force, advanced into Portugal and took Oporto on 29 March 1809 but his subsequent control over the country was very tenuous, due more to the state of the roads and logistics than anything else. There were other French armies under Ney in Galicia and Victor at Merida in Spain, but they were all separated from each other by hostile country and their communications were poor, if even existent. Meanwhile, the British Government had decided to take another hand in Portugal and on 22 April 1809 Wellesley again landed in the country and took over command of the Anglo-Portuguese Army; William Beresford was appointed C-in-C of the Portuguese Army and promised considerable support.

Subsequently, throughout May 1809 they conducted a most successful campaign in Northern Portugal, during which Wellesley made his immortal crossing of the River Duoro, forcing Soult to abandon Oporto and, without fighting a major battle, caused the French Army to lose all its baggage, transport and artillery. Next, the British army marched into Spain to make contact with the Spanish Army under Cuesta, and then attack the French under Victor and Sebastiani south of Madrid, not an easy undertaking owing to the unreliability of the Spanish commander and his forces.

THE BATTLE OF TALAVERA – 27/28 JULY 1809

Opposing forces

Sir Arthur Wellesley and General Cuesta at the head of the Anglo-Spanish Army – 4 British infantry divisions (10 brigades of about 29 battalions of infantry – 16,300 men); 2 brigades of heavy and 1 brigade of light cavalry (7 regiments: 2,969 cavalrymen); 30 guns plus a battery of Spanish 12-pdrs (loaned by Cuesta) 6 guns.
6 Spanish infantry divisions (41 battalions – 28,000); 2 cavalry divisions – 6,000 men; 30 guns.

Marshal Victor, commanding 3 divisions of infantry (33 battalions: 18,280 men); Cavalry brigade (2 regiments – 980 men). General Sebastiani, commanding 3 divisions of infantry (23 battalions – 14,250 men); Cavalry brigade (4 regiments – 1,200 men). Plus the Madrid Infantry Division (9 battalions – 5,150 men) and a Cavalry Reserve of 2 Dragoon divisions (13 regiments – 6,330 men); 80 guns.

The tactical formula that brought continual success to Wellington in the Peninsula was repeated at Talavera when his 20,000-strong British army was attacked by 46,000 French, the ineffectual Spanish allies positioned behind almost impregnable fortification, took little part in the battle. It began with near disaster when a strong French force led by tirailleurs crossed the Alberche River to surprise Donkin's Brigade; at the same time, Wellesley was fired upon and almost captured by light infantry as he was reconnoitring the field from a tower of the nearby Casa de Salinas. The British, Portuguese and King's German Legion infantry were positioned behind (and protected from artillery fire) a low ridge, anchored on the left by a hill the Cero de Medellin. On the foward slopes a screen of riflemen engaged enemy skirmishers and harassed the French columns, steadily advancing upward with shrapnel shells bursting over and amongst them. They advanced in column of divisions, each of 3 battalions with 2 companies formed abreast, 60 men wide and 24 ranks deep so that 4,300 men were in a three-column attack. As the columns reached the crest of the ridge, the riflemen fell back and the double-line of British infantry went forward to the crest, to meet the French with platoon volleys fired at about 50 yards range. In less than three minutes, 10 of these platoon volleys had

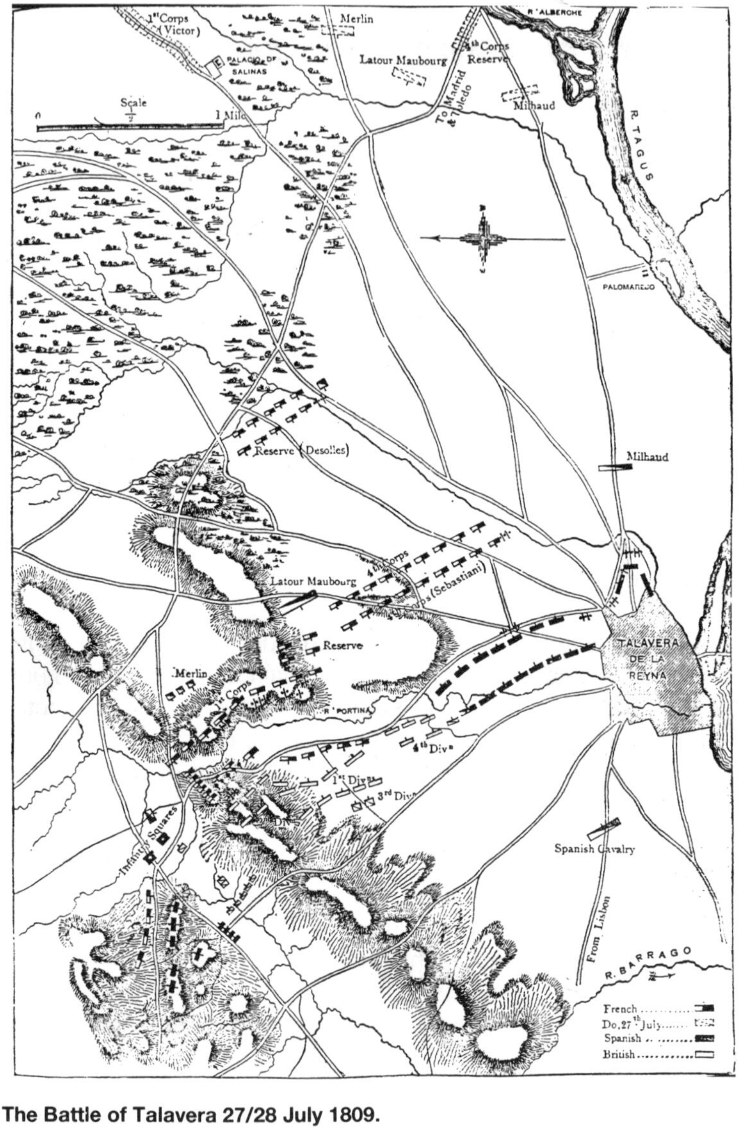

The Battle of Talavera 27/28 July 1809.

Incident in the 48th's famous action at Talavera. The fatally wounded Donellan hands over control: 'Major Middlemore, you will have the honour of leading the 48th to the charge.'

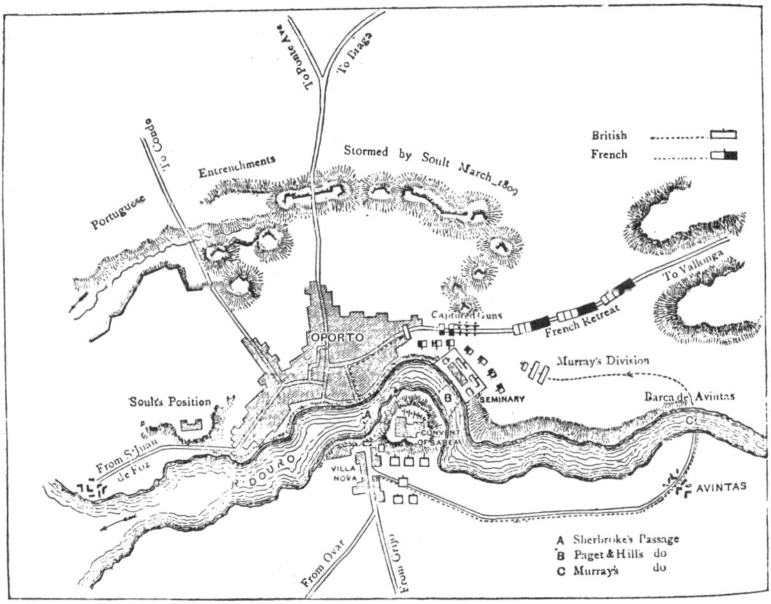

Plan of the passage of the Douro.

been fired before the French broke and ran. Successive column attacks were thrown back, most of their accompanying artillery being captured without firing a shot. Even so, pursuing British infantry (among them Guards and two K.G.L battalions) went out of control and with unloaded muskets came up against overwhelming numbers of French infantry. The British infantry were roughly handled and pursued back to their own lines, where 4 British battalions opened-out, let their comrades through, and then closed to beat back the oncoming French attack. British cavalry also took unnecessary punishment when the 23rd Light Dragoons and the K.G.L Hussars, charging over a plain beyond the British left, hit a shallow ravine at speed, causing losses and so disordering them that the 23rd lost 50 percent on encountering a brigade of French Chasseurs-á-Cheval.

Casualties

Allies – 5,365, almost all British and 25 percent of Wellesley's force.

French – 7,268 (18 percent)

Now followed months of manoeuvring but little fighting, although the over-optimistic Spanish Army (under a new commander, General Areizaga) rashly attempting operations on their own, was soundly beaten at Ocana and Alba de Tormes. Knowing that Spain was temporarily lost, Wellington (as he now was) set about building the Lines of Torres Vedras north and west of Lisbon, and ordered the army out of Spain and back into Portugal. A new French commander, Massena, in direct command of 138,000 troops was ordered to take Portugal, and began to clear his invasion-route by besieging and taking Spanish-garrisoned Cuidad Rodrigo, which Wellington made no attempt to relieve. Next, they laid siege to Almeida, a walled and fortified city just inside the Portuguese border, isolating it after a fierce fight with Craufurd's Light Division, attacked by Ney's entire Corps. Almeida soon capitulated after a massive magazine explosion destroyed much of the place, killing at least 500 of its Portuguese garrison.

THE COMBAT ON THE COA – 23 JULY 1810

Opposing forces

Brigadier-General Robert Craufurd, commanding the Light Division (5 battalions infantry – 1 British Rifles, 2 British Foot; 2

Portuguese Cacadores – 3,780 men); 2 regiments cavalry (776 men); 1 battery horse artillery 6 guns. *Total: 5,492 men*

Marshal Ney, commanding 3 divisions of infantry (30 battalions – 17,750 men); 2 cavalry brigades (4 regiments – 1,680 men); Artillery ? (Napier says 30 guns but this is doubtful). *Total: 24,000 men*

Craufurd's Light Division had been engaged in daily skirmishes with the French for some weeks and, confident in his superbly trained force, he rashly lingered in the exposed area between Almeida and River Coa, with the river and only a single narrow bridge to his rear. Before dawn French cavalry beat in Craufurd's piquets, then Ney sent forward his entire army of 24,000 men in one deep broad column, the cavalry brigades in front, then 13 battalions of Loison's Division in line of columns, then Mermet with 11 more, with Marchand's 3 battalions forming the reserve. Against this host, Craufurd drew up his force over a front of 1½ miles, convex in form, with its left resting on a ruined windmill 500/700 yards south of Almeida (still in Allied hands). From left

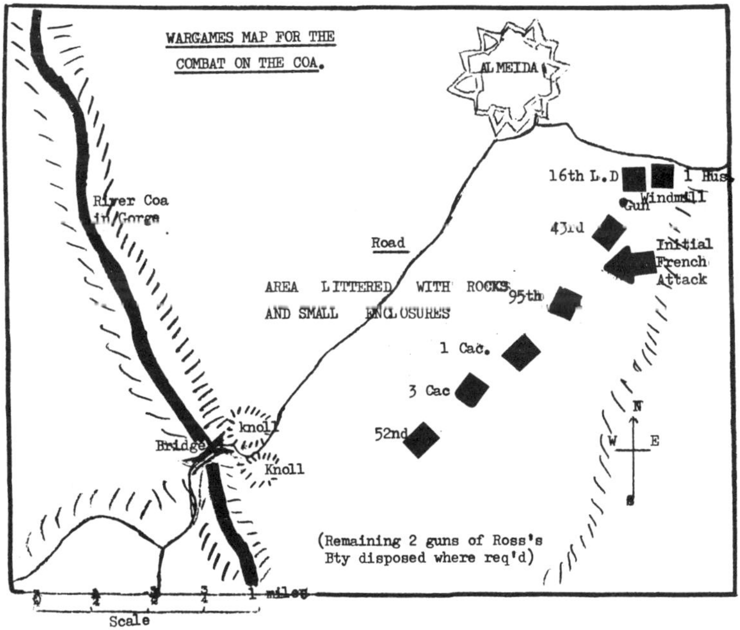

Wargames map for the Combat on the Coa.

Major-General Robert Craufurd. (Born 1764 Died 1812)

to right – ½ company 52nd with 2 of Ross's guns; then the 43rd, then in succession the 95th, 1st Cacadores, 3rd Cacadores and the rest of the 52nd; their right was not far from the river. A classic combat ensued over a steep slope intersected with stone walls and enclosures, beginning on the British left and spreading over the whole area, with cavalry running loose sabreing the small groups of British and Portuguese fighting desperately towards the bridge. Covered by companies of the 43rd and 95th, holding two knolls above the river, Craufurd's force reached the other side, hindered by an overturned limber on the many-angled bridge. Finally all were over and in strong positions amid the rocks, holding off with deadly fire the frantic attacks of infantry sent over the narrow bridge by Ney, until it was choked to its parapets with dead. The fight petered out by four in the afternoon, amid torrents of rain.

Casualties

British – 36 officers and men killed; 214 wounded; 83 missing – Total: 333 men.

French – 520 killed and wounded.

Almeida. Taken from the top of the main gatehouse, showing fortifications and ditches.

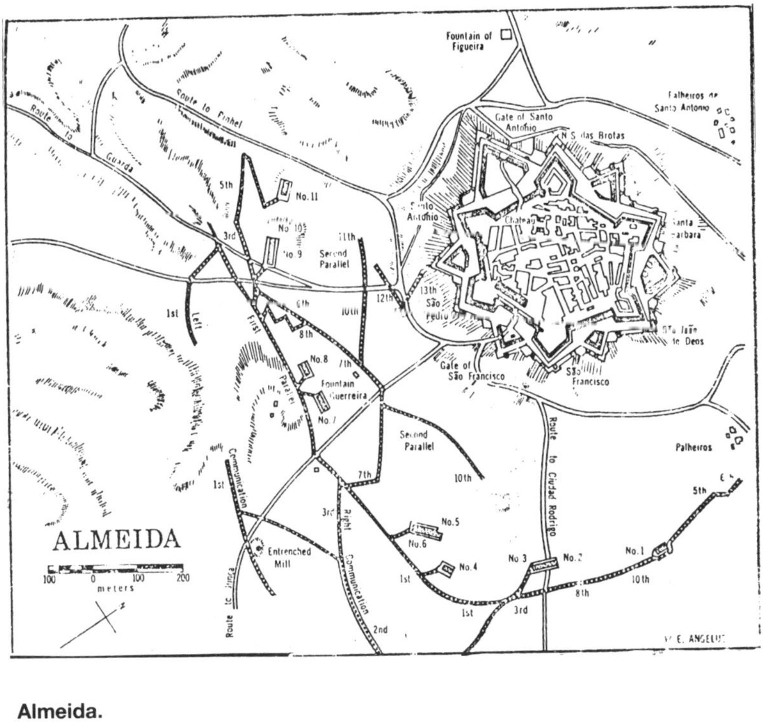

Almeida.

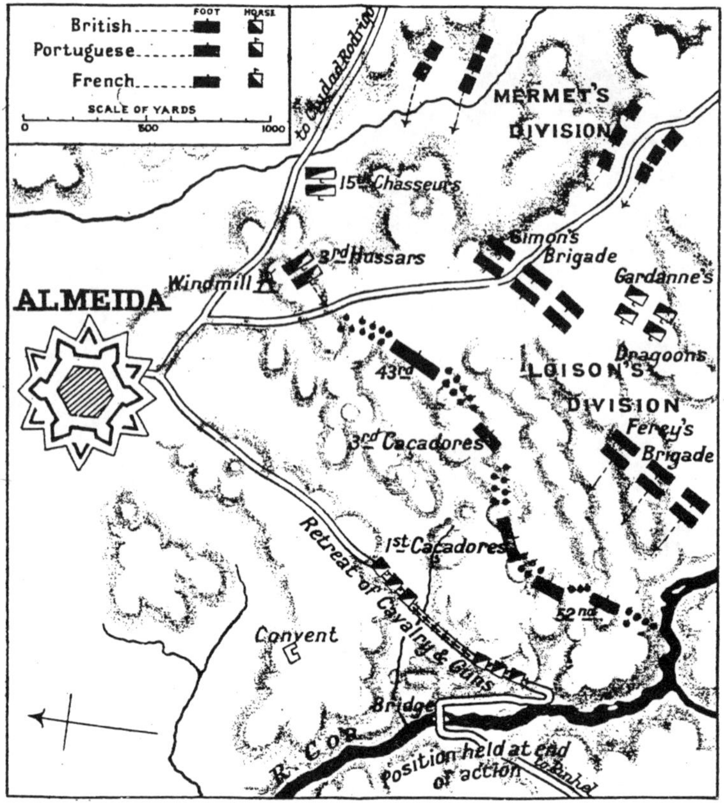

Combat of the Coa, 24 July 1810.

Keeping to the route north of the Mondego River, Massena marched south through country cut by narrow defiles, where Portuguese guerillas maintained day and night sorties. Wellington concentrated all his forces on the Busaco Ridge, north of the Mondego River and athwart the highway to Coimbra; after 14 months he was ready and willing to give battle.

THE BATTLE OF BUSACO – 27 SEPTEMBER 1810

Opposing forces

General Viscount Wellington, commanding the Allied Army of 5 British infantry divisions; the Light Division; 1 Portuguese division and 3 independent brigades (85 battalions – 24,796 British; 24,649 Portuguese men); 2 squadrons cavalry – (210 men); 60 guns.

Total: 51,345 men

The Battlefield of Busaco. View of Sula across the valley, showing slopes up which Loison and Marcune attacked towards Sula and then up to the heights above the village. The heights on the top left wooded skyline are where Craufurd was positioned. This picture gives a good impression of the very difficult terrain, although it was far less wooded then than now.

Marshal Massena, commanding L'Armée de Portugal of 3 Corps d'armée (7 divisions; 15 brigades; 84 battalions – 50,733 men); 1 cavalry division and 5 brigades (11 regiments and 12 squadrons – 7,612 men); 114 guns. *Total: 65,974 men*

The Sierra de Busaco extended 10 miles, rose almost 500 feet in places and was so precipitous that there was no area where it could be climbed easily, nor any route where it could be approached unseen – if the French wished to proceed then they would have to push 51,000 men from their positions. Wellington using the terrain with great skill to conceal his men, posted his infantry amid the gorse and heather along the top of the ridge, siting his 60 guns amid them in small numbers, to be used at close range. As usual, the French attacked in columns, laboriously picking their way upwards behind screens of tirailleurs; in the same old way British and Portuguese riflemen on the forward slopes poured in their fire, scurrying from boulder to boulder as they fell back towards the summit of the ridge. Carrying 60lbs of equipment per man, struggling upwards in closely packed columns 40 men wide and 50 men deep, harassed by the skirmishers and then by rolling-volleys from ranks of experienced infantry and showers of grape from cannon, the severely suffering

French infantry were brought to a halt, and then sent tumbling down the steep slopes. Two columns climbed the northern face of the ridge where seemingly there were no defenders, then suddenly Craufurd appeared alone before them on the crest, followed by 1,800 redcoats rising from concealment like devils at a pantomime. A flaming line of fire rippled along the line to bring down scores of French infantry, then the British moved slowly

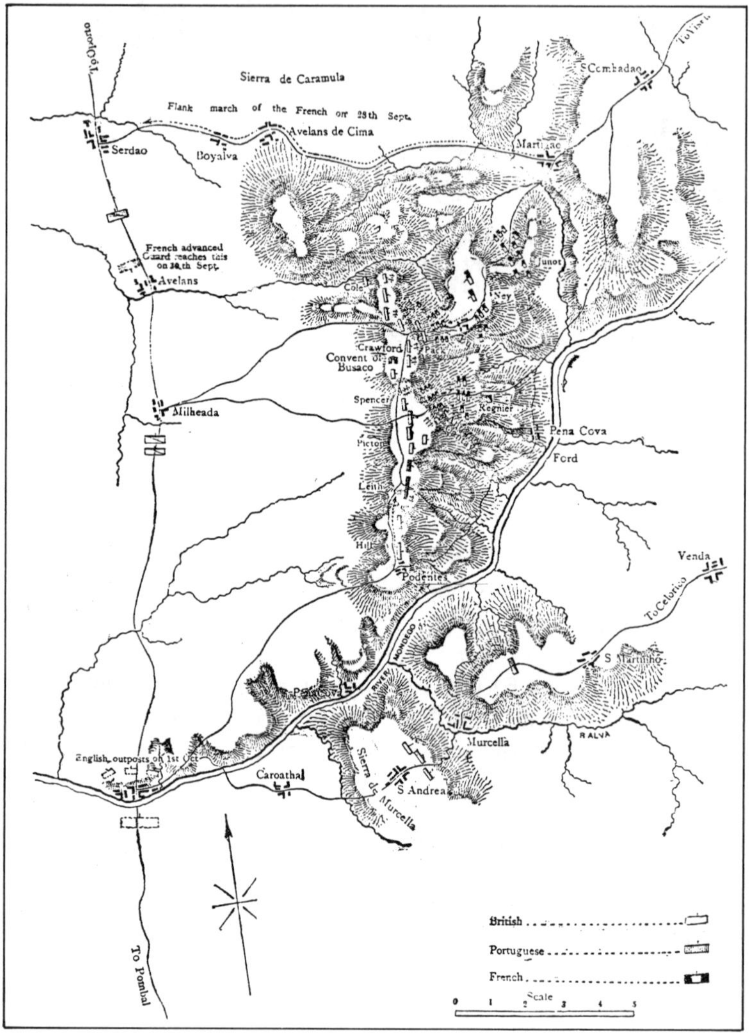

The Battle of Busaco 27 September 1810.

forward pouring in volleys as they advanced; demoralised and with most of their officers down, the sight of the long gleaming bayonets was the last straw. Disintegrating into a disorderly mob they were bundled unceremoniously down the slopes and through the narrow streets of the village of Sula amid showers of grape from the artillery above. By mid-afternoon Massena, realising that he had launched his attack without proper reconnaissance when he should have tried to turn the position, abandoned the attack.

Casualties

British – 544 British and 626 Portuguese killed and wounded.
French – 4,473 killed and wounded.

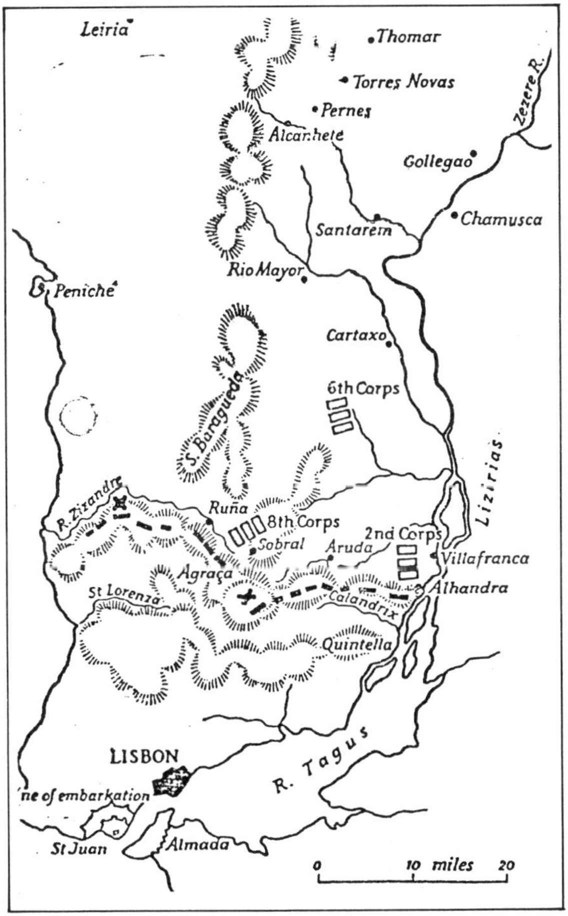

Torres Vedras.

However, the next day Massena discovered a way around the Busaco Ridge and his subsequent outflanking move compelled Wellington to evacuate the position and withdraw to the Lines of Torres Vedras, constructed during the winter of 1809/10. They consisted of a series of redoubts and fortified positions amid the mountainous country stretching from Lisbon to Torres Vedras, noted for its abrupt peaks, deep gullies and ravines; stretching for 30 miles from the Tagus River to the sea, they mounted some 600 guns, mostly supplied from Portuguese arsenals and mounted on naval carriages so that they could not be hitched up and taken away to be turned against the Allies, should Massena force the Lines.

By mid-October, Wellington was firmly installed within the Lines. Massena, to whom these defences had come as a great shock, fruitlessly probed them and then settled down to blockading the area, living precariously in a devastated inhospitable country where, by the end of the winter, 25,000 of his men had been lost through wounds, disease and malnutrition. Early in March 1811 the French withdrew into Spain, harassed by Wellington and guerillas, reaching their base at Salamanca in early April, after Massena had skilfully extricated his army from a number of critical situations. Within a month, however, Massena was again on the march, attempting with nearly 50,000 men to relieve Almeida when he came up against Wellington's army in the north, between the rivers Dos Casas, Turones and Coa in an area astride today's Portuguese/Spanish frontier.

The Lines of Torres Vedras. St. Vincente Fort S.W. of town of Torres Vedras.

THE BATTLE OF FUENTES D'ONORO – 3 AND 5 MAY 1811

Opposing forces

Lord Wellington, commanding the Anglo-Portuguese Army, consisting of 6 infantry divisions and 1 independent brigade (18 brigades; 66 battalions – 23,026 British; 11,471 Portuguese); 3 cavalry brigades (6 regiments – 1,542 British; 312 Portuguese); 48 guns. *Total: 37,614*

Marshal Massena, commanding 4 corps (8 divisions; 11 brigades; 83 battalions – 42,206 men); 6 cavalry brigades (19 regiments – 4,662 men); 38 guns. *Total: 48,268*

On 3 May, Massena came up with Wellington's army positioned behind the village of Fuentes d'Onoro, and two days of hard fighting followed with many desperate hand-to-hand encounters within the walled confines of the village, where the light companies and riflemen showed more aptitude for street fighting than the French, accustomed to attacking in column. Generally, the British/Portuguese open-order fighting was far more effective than the French mass system. At one stage the Light Division with Cotton's cavalry found themselves isolated on an open plain surrounded by large numbers of French cavalry, plus 3 infantry divisions and several artillery batteries. Craufurd handled his command with great skill, keeping the line-battalions in mobile squares while employing his riflemen as skirmishers, not allowing himself to be pinned down by enemy cavalry which would allow their infantry and artillery to pour fire into his formation. Slowly retiring, the riflemen using the cover of low rock outcrops, maintained a galling fire on the enemy at about 250 yards range, running to the cover of squares when cavalry approached; on those occasions when French artillery came near, they were forced back by the threat of being charged by Cotton's cavalry. With the squares moving alternately back, Bull's Horse Artillery positioned themselves between them, keeping off the approaching enemy. In their two mile withdrawal the Light Division, showing the highest discipline, courage and confidence, lost only 35 men, while simultaneously a British Army epic occurred when Captain Norman Ramsay's two guns of Bull's Troop, completely enveloped by French cavalry, cut their way through the encircling mass at the gallop, the mounted gunners surrounding the limbered pieces and caissons.

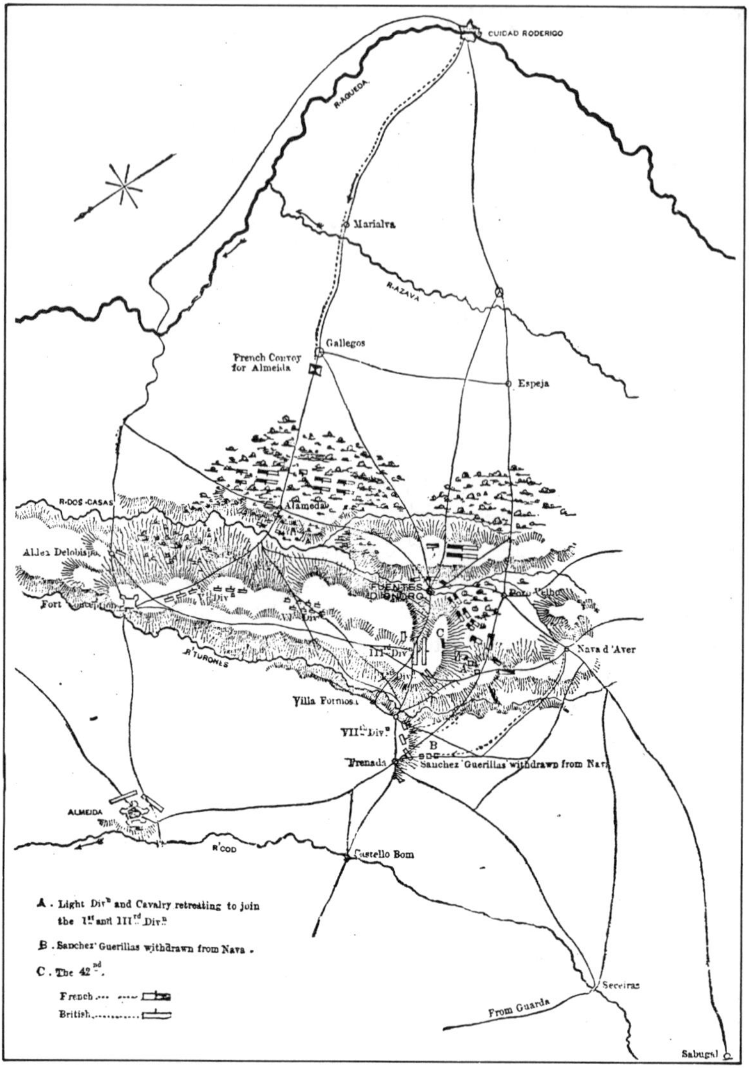

The Battle of Fuentes d'Onoro 3 & 5 May 1811.

Casualties

Allies: 1,545 French: 2,192

Abandoning his intentions of relieving Almeida, Massena fell back east of the Agueda on 10 May.

As long as the French retained their grip on the vital fortress towns of Almeida, Badajoz and Cuidad Rodrigo (all on or near the frontier between Portugal and Spain) Wellington could not safely venture across the central and northern parts of Spain to attack the enemy's lines-of-communication with France. To do so would leave exposed his own links with the south and his base at Lisbon. Late in April 1811, part of the army under Marshal Beresford laid siege to Badajoz; however, his artillery resources were deplorably weak with many of the guns obsolete, and there was a lack of proper siege materials and trained engineers. But before the siege really got under way, Beresford heard that Soult, with 18,000 troops, was marching to relieve the fortress-town, which forced him to raise the siege and march out to meet the enemy.

THE BATTLE OF ALBUERA – 16 MAY 1811

Opposing forces

Marshal William Beresford, commanding an Anglo-Portuguese Army, in association with a Spanish Army under Generals Blake and Castanos. Three Anglo-Portuguese infantry divisions and 2 independent brigades (9 brigades and detached light companies of 3 regiments – 32 battalions – 8,738 British; 9,131 Portuguese); 3 brigades cavalry (5½ regiments – 1,416 British; 849 Portuguese); 36 guns. *Spanish* – 3 infantry divisions and 1 brigade (12,593 men); 2 brigades cavalry (1,866 men); 2 batteries artillery.
Totals: Anglo-Portuguese: 20,310 Spanish: 14,644

Marshal Soult, commanding 2 divisions and 3 brigades of infantry (34 battalions – 19,015 men); 3 brigades and 3 unattached regiments of cavalry (12 regiments in all – 4,012 men); 48 guns.
Total: 24,260

At Albuera, Soult tactically outmanoeuvred Beresford (but would probably have done the same to any other Allied commander in the Peninsula, save Wellington himself) who was handicapped by his soldiers not having the same confidence in him, despite his being brave, loyal and talented. After early skirmishing, the French made the most massive single attack of the whole war, when 8,500 men – two full infantry divisions – in an almost solid column with only a light skirmish line preceding it, artillery in support, and left flank protected by 3,500 cavalry, attacked four battalions of Spanish infantry on the Allied southern flank. Fighting unusually well, the Spanish would have been

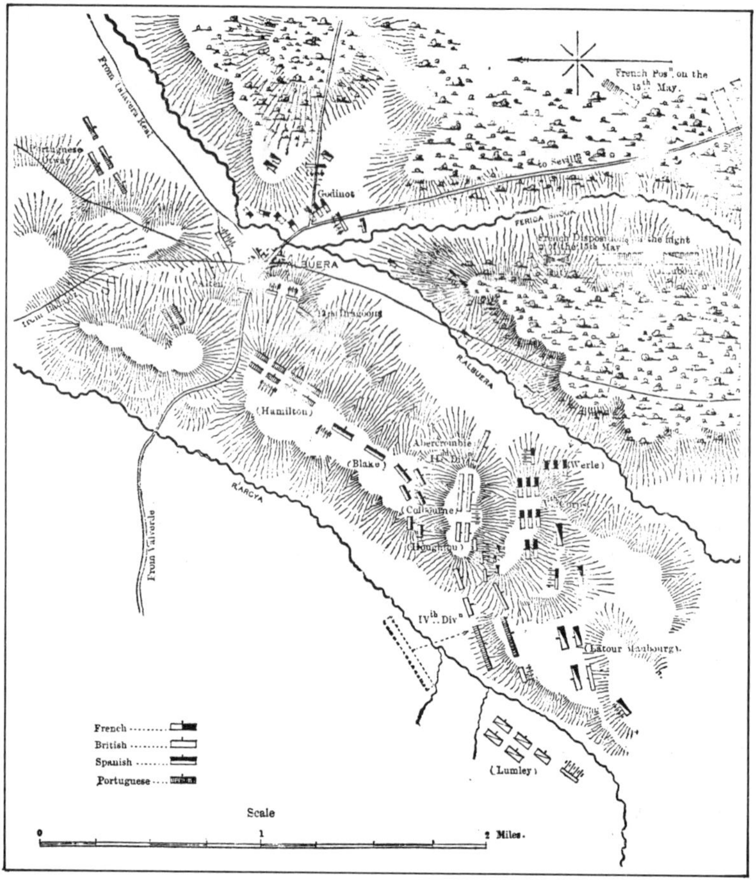

Plan of the Battle of Albuera.

broken had not a British division come up on their right flank, while Colborne's brigade attacked the flank of the column. A sudden storm of rain caused muskets to be put out of action, allowing two regiments of Polish lancers to approach unseen and hit one end of the British line, the lances of the experienced horsemen in tight formation practically annihilated the first three of Colborne's battalions, before breaking into the rear of the Spaniards to sweep over a battery of K.G.L. guns and almost catch Beresford and his staff. Rallying, the Spanish infantry returned, reinforced by 2 British brigades and the remaining battalion from Colborne's stricken brigade; muskets could now be fired as the rain had ceased. In a two-deep line, 7 British battalions of 3,700 men now began a close-range firefight with

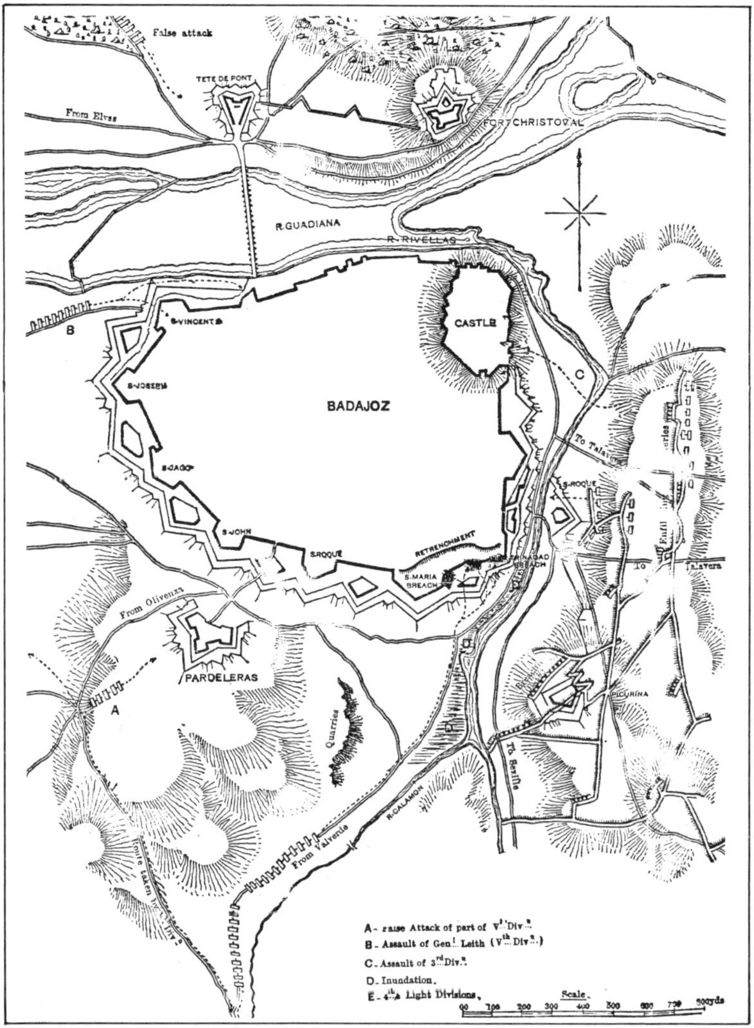

Plan of the Assault of Badajoz.

the huge French column, 200 men wide and 40 ranks deep. Over an area about the size of a cricket field, this contest dragged on for nearly an hour; several French batteries only 300 yards behind the head of the column, enfiladed and cross-fired through the centre of the British line. These showers of grape and canister would have caused astronomical losses but for targets being obscured by smoke and heavy cloud. Torrents of rain brought the battle to an end with the armies remaining in position until late on the following day, when the French retired covered by their horse artillery and cavalry.

Casualties

British 4,199; Portuguese 389; Spanish 1,368. French 6,000/8,000 killed, wounded and missing.

The French onslaught on Portugal had failed, the victories at Fuentes d'Onoro and Albuera having established Allied supremacy over both routes between Spain and Portugal; but the French still had five times as many men in the Peninsula than Wellington could assemble and every Spanish army taking the field had been defeated. Wellington could not undertake an offensive until he had recaptured Ciudad Rodrigo and Badajoz, from either of which the French could advance again. In a short, sharp siege lasting only from 8 to 19 January 1812, Cuidad Rodrigo was stormed relatively inexpensively, although Craufurd was killed in the assault. However, Badajoz was a much more difficult proposition, the siege lasting from 16 March to 6 April 1812, and the storming was one of the most costly Allied

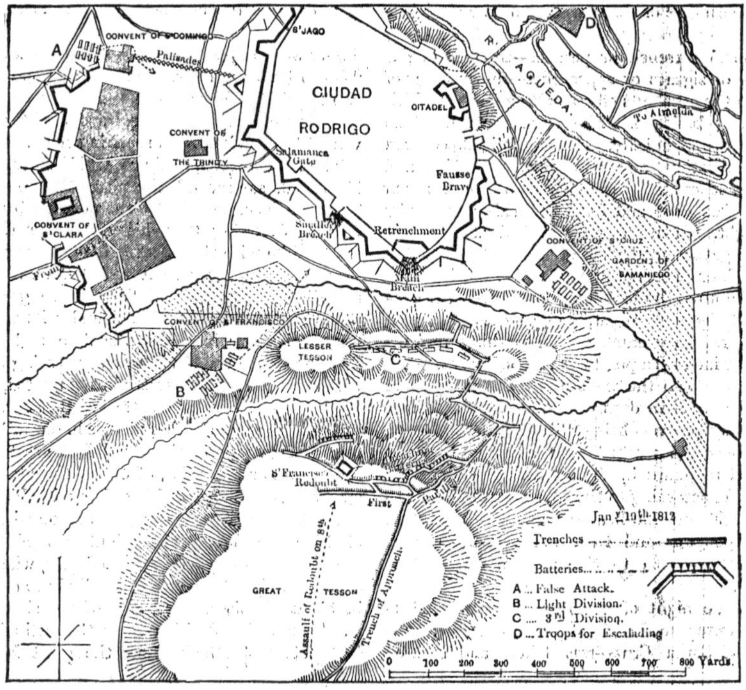

Plan of Assault.

operations of the whole war, more men being lost than in any other major battle, with the exceptions of Talavera and Albuera.

Wellington was now able to thrust boldly into Northern Spain, with Marmont who had taken over from Massena, attempting to sever his communications with Cuidad Rodrigo; in mid-July the two armies marched parallel with each other for several days, each commander keeping his army well in hand as he sought a tactical advantage. Eventually, Wellington seized a favourable opportunity to engage Marmont, the resulting battle being a classic engagement, probably Wellington's masterpiece.

THE BATTLE OF SALAMANCA – 22 JULY 1812

Opposing forces

General the Earl of Wellington, commanding an Anglo-Portuguese Army, formed of 7 infantry divisions (22 brigades; 92 battalions – 25,577 British; 17,421 Portuguese – 42,998 men); 5 brigades cavalry (12 regiments –3,543 British; 482 Portuguese); 54 guns (British) 6 guns (Portuguese). *Total: 48,569*

Marshal Marmont, commanding L'Armèe de Portugal, composed of 8 infantry divisions (16 brigades; 75 battalions – 41,575 men); 2 cavalry divisions (26 squadrons – 3,390 men); 131 batteries of 78 guns. *Total: 49,999*

The Battlefield of Salamanca. The Greater Arapile taken from the top of the British-held Lesser Arapile. Much fighting over this ground.

At midday on 22 July, Marmont had extended his left flank allowing Pakenham's 3rd Division with Portuguese cavalry to advance under cover of a range of hills and drive into the leading French column (Thomiere's) before they knew what had hit them. Clausel's 2nd infantry division were next hit by three volleys of a two-deep infantry line, the survivors breaking and fleeing, the disintegrating force being charged by Le Marchant's Heavy Cavalry, who smashed five battalions. Then they charged into Ferrey's 3rd Division, exhausted from double-marching for a mile and, although in squares, were unable to withstand the British Dragoons and were scattered. A French counterstroke of 12,000 men – in columns lacking supporting artillery and inferior numbers of riflemen – was stopped in its tracks and completely broken in five minutes. Of the original 8 French divisions, five were now dispersed and both Marmont and General Bonnet were casualties; Clausel had taken over command and, for the first time in the Peninsula, formed the 7 battalions of a complete French division into a continuous three-deep line with a single battalion – in square – at either end; but it was quickly broken by concentrated artillery fire. The French army was now in full retreat but the victors, physically exhausted, were unable to pursue.

Major-General Le Marchant, founder of the Royal Military Academy, Sandhurst. He was killed leading the Dragoons at Salamanca.

Casualties

British 3,176; Portuguese 2,831. French at least 14,000 killed, wounded and missing. The French army was almost completely lost or dispersed.

On the following day the cavalry brigades of Anson and Bock came up with the French rearguard at Garcia Hernandez, where ensued a most spirited action between 4 squadrons of K.G.L

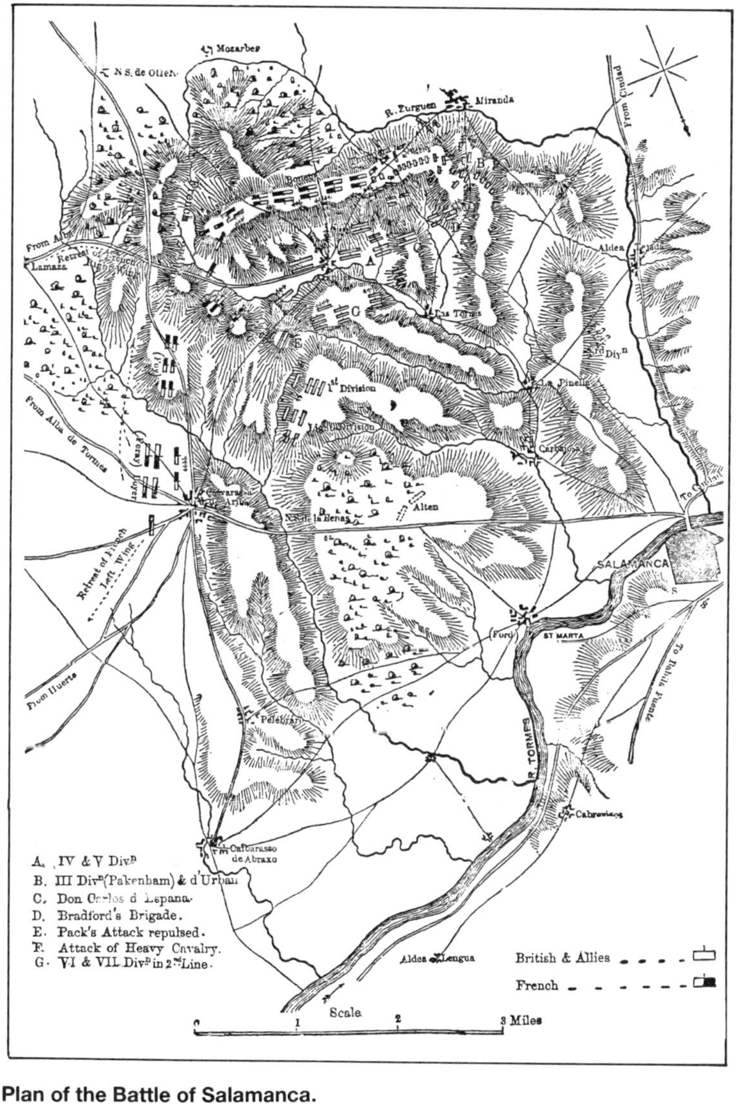

Plan of the Battle of Salamanca.

heavy dragoons and 3 battalions of French infantry, where the cavalry broke a square and scattered a 2-battalion-strong column.

After marching to and entering Madrid on 12 August, Wellington decided he could make no better use of his army than to drive Clausel further along the road that led back to France; but the old castle at Burgos, well garrisoned and supplied, lay in the way. It was stronger than Wellington thought, and for a month fruitless siege operations made little impression on the place until, on 9 October, the news came that the armies of Soult, Joseph and Suchet had combined and would probably be too strong for Hill, in Madrid with an Allied army of 36,000. Both Wellington and Hill began retreating north and, after many skirmishes particularly at river-crossings, were united at Cuidad Rodrigo by late November, after a brilliantly carried out withdrawal notable for Soult's refusal to allow 100,000 Frenchmen to come to grips with 52,000 Anglo-Portuguese. The army wintered in the area and, after being steadily reinforced throughout the period, began operations at the end of May 1813.

Abandoning Madrid, the French began retreating northwards hustled along by the Allies; the various dispositions of French armies throughout Spain meant that Wellington had a reasonable numerical superiority over the forces of King Joseph and Marshal Jourdan before him who, by mid-June, had their forces positioned behind the Zadorra River covering Vittoria. Here, Wellington decided to attack them.

THE BATTLE OF VITTORIA – 21 JUNE 1813

Opposing forces

The Duke of Wellington, commanding an Anglo-Portuguese and Spanish force consisting of 8 infantry divisions and 2 independent brigades (24 brigades; 81 battalions – 27,372 British; 27,569 Portuguese men); 8 British and 2 Portuguese brigades of cavalry (19 regiments – 7,424 British; 893 Portuguese); 78 British guns; 12 Portuguese guns. *Spanish* – 1 infantry division and 1 brigade of 5 battalions (6,800 men). *Total – 79,062* – made up of 37,800 British; 28,762 Portuguese; 7,000 Spanish.

King Joseph of Spain and Marshal Jourdan, commanding 8½ infantry divisions (17 brigades; 35 regiments – 49,000 men); 6 cavalry divisions and 1 brigade (29 regiments – 11,300 men); 138 guns. *Total: 66,000*

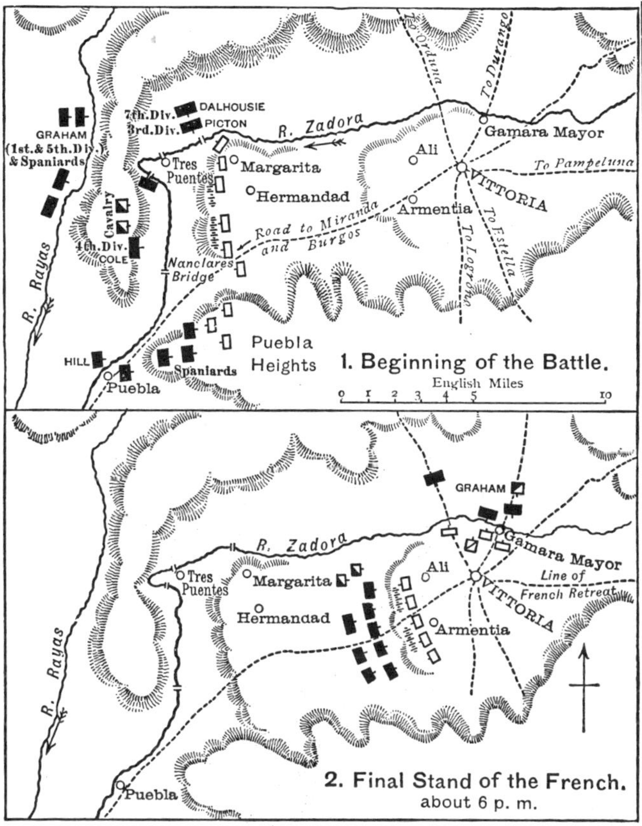

Wellington's Army **Battle of VITTORIA.** French Army

Expecting a straightforward frontal attack from the west, Joseph and Jourdan disposed accordingly, with Gazan's 4½ divisions facing Wellington south and west of the Hill of Arinez, supported by guns on the hills to their rear; D'Erlon's 2 divisions were behind them, drawn up east of Arinez on either side of the Great Road; Reille's troops formed a third line just west of Vittoria itself. Wellington had no intention of attacking frontally and split his army into four – Hill led 20,000 men over the hills and Pass of Pueblo, to drive in on Gazan's left; Picton's 3rd and Dalhousie's 7th divisions were to enter the valley from the north-west to fall on Gazan's right and rear; Wellington himself led the Light and 4th Divisions in a frontal attack; and finally, Graham would take 25,000 men in a huge northward swing, to come in on the north-east corner of the valley, engage Reille and cut the Great Road to Bayonne. Amid fierce fighting, the plan went as

The Battlefield of Vittoria. The bridge of Nanclares over which Wellington led Lowry Cole's 4th Division. In the background are the Heights of Pueblo over which Hill's force attacked.

hoped and by afternoon the French decided to abandon the valley and began withdrawing in orderly fashion towards Vittoria, then Graham's force arrived on the field. Hearing his brisk volleys behind them and astride the road to Franch did not help French morale, formations began to break and soon the entire army was streaming in panic-stricken rout from the field, having to take the Pamplona road because Graham had cut the road to Bayonne. In this decisive battle of the Peninsular War, the French left the battlefield in an unparalleled state of litter and confusion, abandoning everything including all their Spanish booty and every gun except two.

Casualties
British: 3,675 killed, wounded and missing. Portuguese 921; Spanish 552. French: 8,000.

Joseph retired across the Pyrenees unpursued by the Allies, who were aware that Napoleon himself was back in France –

having signed an armistice with Russia and Prussia – and might well bring an army of overwhelming strength against them. Marshal Soult replaced Jourdan and reorganised the French army, still stronger than the British who had less than 60,000 Anglo-Portuguese troops over a front of 40 miles; they were positioned in depth to take advantage of the road system. Pamplona and San Sebastian went under siege and, on hearing of an unsuccessful attack on the latter on 25 July, Wellington rode over there from his H.Q. at Lesaca – while he was away Soult struck.

THE BATTLE OF RONCESVALLES – 25 JULY 1813

Opposing forces

Major-General Sir Lowry Cole, commanding 1 British division and 1 brigade of Infantry (4 brigades; 13½ battalions (3 Portuguese)

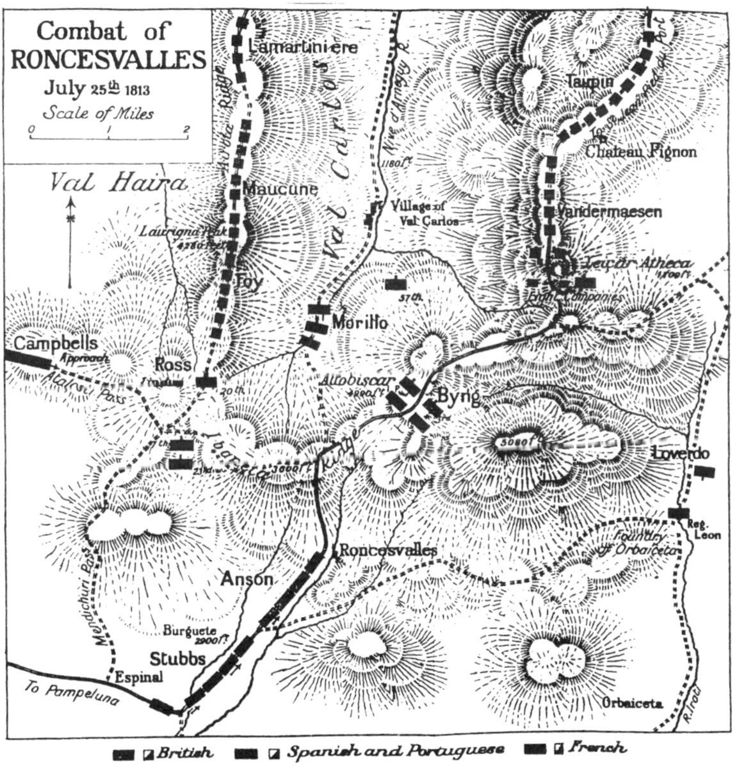

Combat of Roncesvalles.

– 10,000 men); Campbell's Portuguese infantry brigade (3 battalions – 2,795 men). Number of guns not known. *Total: 13,500*

Marshal Soult, commanding 6 infantry divisions (12 brigades – 34,453 men). Number of guns not ascertainable.

The French objective was to force the passes at Maya and Roncesvalles, the outposts at the latter pass were not surprised and they fell back in good order, delaying the advance of the French column stretched out for miles over the narrow approach road. Cole held off the continual attacks by skilful use of the ground; at one point 500 men of British light companies deployed on a front of 300 yards with deep valleys on either side, held for four hours. Despite being forced back by sheer weight of

The Battlefield of Roncesvalles. There is a narrow track running down the hillside (just above centre white patch). Down it marched Reille's three French infantry divisions, advancing on the Linduz Peak west of Roncesvalles. The track enters the centre trees, where the fighting took place.

So narrow was this track that they marched two-abreast and when leading troops were fighting the British, the column's rear was 11 miles back!

numbers, the British force never lost control of the pass and were still holding it when, in late afternoon, a thick fog brought fighting to an end. However, notwithstanding his successful defence, Cole became nervous and ordered a retreat along the road back to Pamplona.

Casualties

British – approx. 345 killed, wounded and missing.
French – approx. 200 killed, wounded and missing.

An artist's idea of how the fighting between the 20th Regiment and Reille's infantry must have looked at Roncesvalles.

THE BATTLE OF MAYA – 25 JULY 1813

Opposing forces

Lieut. General William Stewart, commanding Cameron's 1st and Pringle's 3rd brigades of Stewart's 2nd division, plus 3 battalions (Barnes' brigade) of the 7th division (9½ battalions – 6,000 men). Number of guns not ascertainable.

The Battlefield of Maya. The heather-covered slope of Mont Alcorunz stretching right, up from the light coloured area on the left, is Chemin des Anglais where a wing of the 92nd Regiment held off a French Division.

General D'Erlon, commanding 3 infantry divisions and auxiliaries (6 brigades 20,000 men). Number of guns not ascertainable.

The placing and supervision of the British advance posts was faulty and the sudden French attack pushed them in with heavy loss, and reinforcements were fed-in piecemeal and not to best effect. Systematically, the French drove back the British infantry, except at one point where ½-battalion 1/92nd (Gordon Highlanders) in a two-deep line on a 50-yard-wide crest, fought an entire division for twenty minutes. With 4 British battalions opposed to 3 divisions, the valley was taken, but was regained in early evening when two fresh British battalions reached the field, drove into the French flank and pushed them back into the Pass. After ten hours of very bloody fighting, the action ended when darkness obscured all targets. Four Portuguese field-pieces were captured, the only guns ever taken from Wellington in any action. The French ultimately secured the Pass, when Stewart was ordered to fall back to Elizondo.

Casualties

British – 1,347 killed, wounded and missing.
French – 2,000 killed, wounded and missing.

Falling back, the Allies halted on ridges barely three miles from Pamplona, but Cole, with foresight, persuaded Picton to allow him to take position on a ridge above the village of Sorauren. Returning to the area, Wellington ordered all withdrawing troops to concentrate on the position, before dramatically joining them, a solitary horseman, with pursuing French patrols almost on his heels.

THE BATTLE OF SORAUREN – 28 JULY 1813

Opposing forces

The Duke of Wellington, in command of part of the Anglo-Portuguese Army, formed of 3 infantry divisions (9 brigades; 28 battalions – 24,000 (the action was mainly fought by about 12,000 men of Cole's, Pack's and Morillo's Spanish divisions). Number of guns not ascertainable.

Marshal Soult, commanding 6 infantry divisions (12 brigades; 24 regiments – 36,000 men – the action was mainly fought by about 25,000 men). Number of guns not ascertainable.

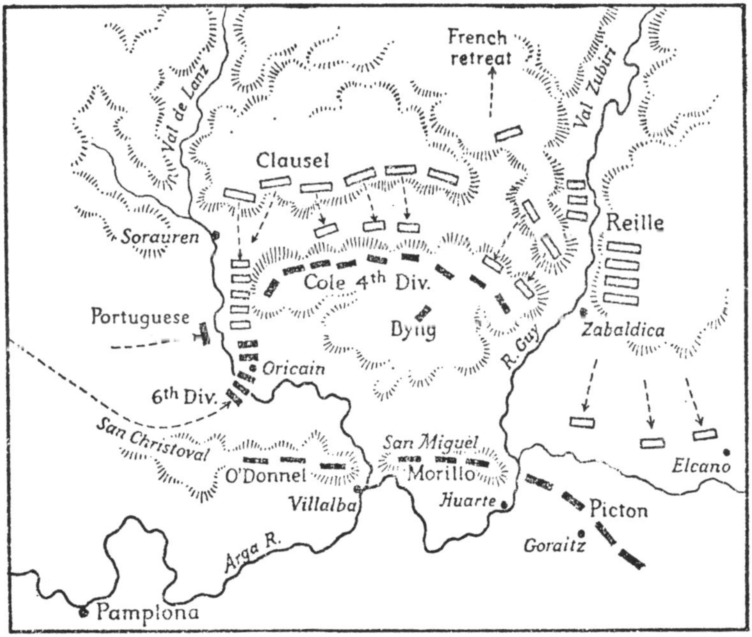

The Battle of Sorauren.

Cole's division of 4 British and 1 Portuguese brigades was on a ridge facing Soult's army; Pack's 6th Division had arrived in the valley on their left; the right was held by Spanish troops on a hill to the rear, Pack repulsed an attack by Conroux's division, while four more French divisions attacked Cole, who held on the right but gave ground on his left, which was regained by a spirited counter-attack. Morillo's Spaniards were almost overwhelmed by superior numbers, but the position was held by a British battalion in line on the crest. The French were driven back into the frontal valley and fighting petered out, with the Allies still firmly in control of the Pamplona road. Two days later on the 30th, an Allied attack destroyed the French in the area, only Foy managed to get his division back to France via goat-tracks over the hills.

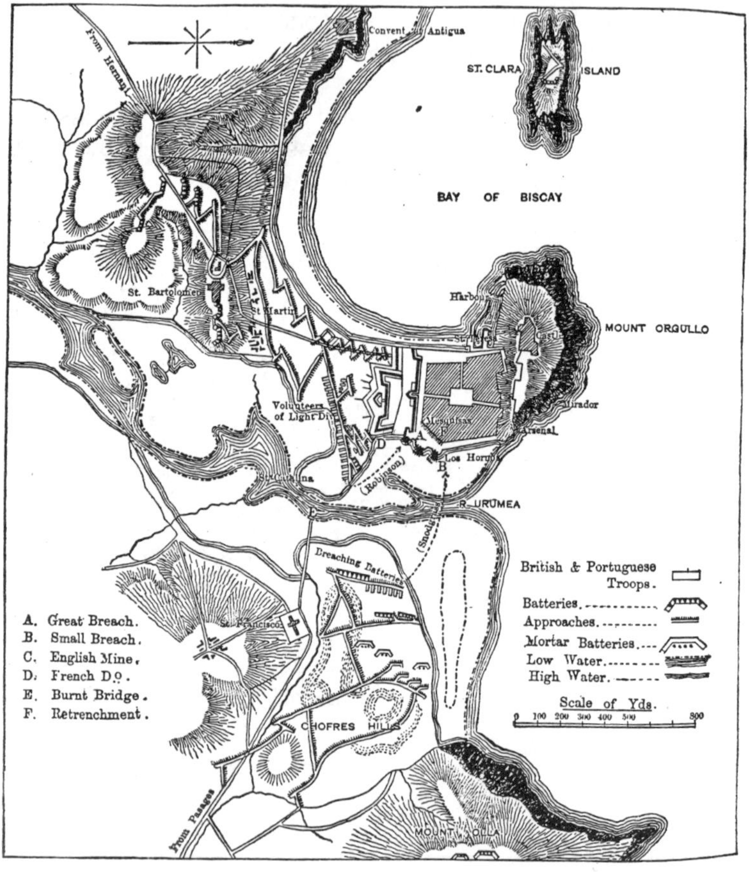

Plan of the Assault of St. Sebastian.

Wellington did not press his full advantage, content at that stage to take San Sebastian and Pamplona.

Casualties

Allies – 2,652 (1,358 British; 1,102 Portuguese; 192 Spanish).
French – approximately 3,000–4,000 killed, wounded and missing.

San Sebastian was an immensely difficult place to besiege, with the sea washing its walls on most sides; an assault failed on 25 July and it was not until 31 August that Wellington was able to fully attack. The 5th division with Bradford's Portuguese and volunteers from 1st, 4th, and Light divisions did most of the attacking, although two Portuguese infantry battalions made a gallant secondary attack wading over 700 yards of mud-flats and deep-water pools in the estuary. Held up in the breaches, the way was cleared when Graham ordered accurate artillery fire just over the heads of his stormers. The town fell, and the castle itself was battered into submission on 5 September.

At the end of August, Soult made two attempts to restore the situation, crossing the Bidassoa River in strength he was thrown

San Sebastian.

back at San Marcial by an all-Spanish force under General Freire. On the same day he sent four divisions against Vera, but the half-hearted attempt was easily repulsed; a sudden storm caused the river to rise and Vandermaesen's division had an epic struggle to cross the bridge at Vera and reach safety, being held for two hours by Captain Dan Cadoux and 85 riflemen of the 95th.

Wellington was ready to strike by early October, his objective being the great spine of the Pyrenees and the Greater Rhune, on the borders of France.

THE BATTLE OF VERA – 17 OCTOBER 1813

Opposing forces

The Duke of Wellington commanding part of the Anglo-Portuguese Army comprising the light division (2 brigades; 8 battalions – 2 Portuguese – 5,484 men); Longa's Spanish infantry division (2,607 men); Giron's Spanish divisions (2) – (7,653 men). Number of guns not known. *Total: 15,800 men*

Marshal Soult, commanding Clausel's Corps of 3 divisions of infantry (27,300 men). Number of guns not known. (The total

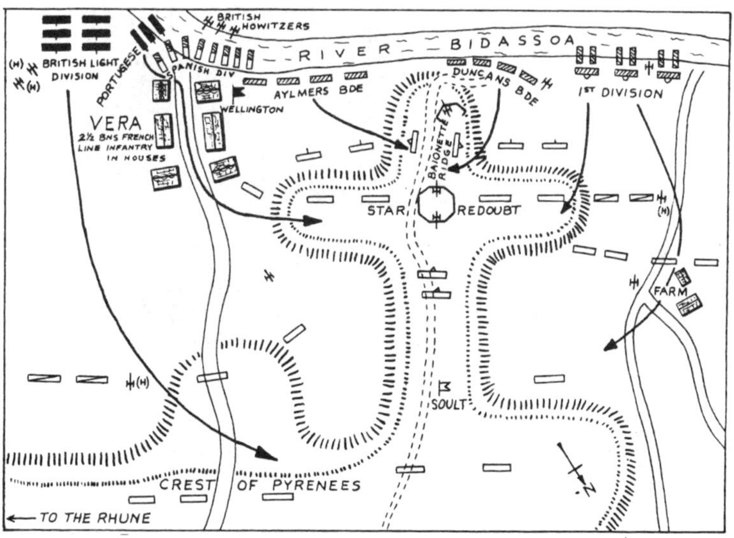

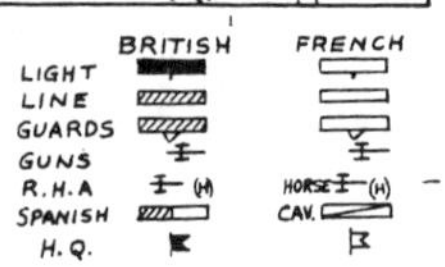

Battle of Vera (fought as a Wargame).

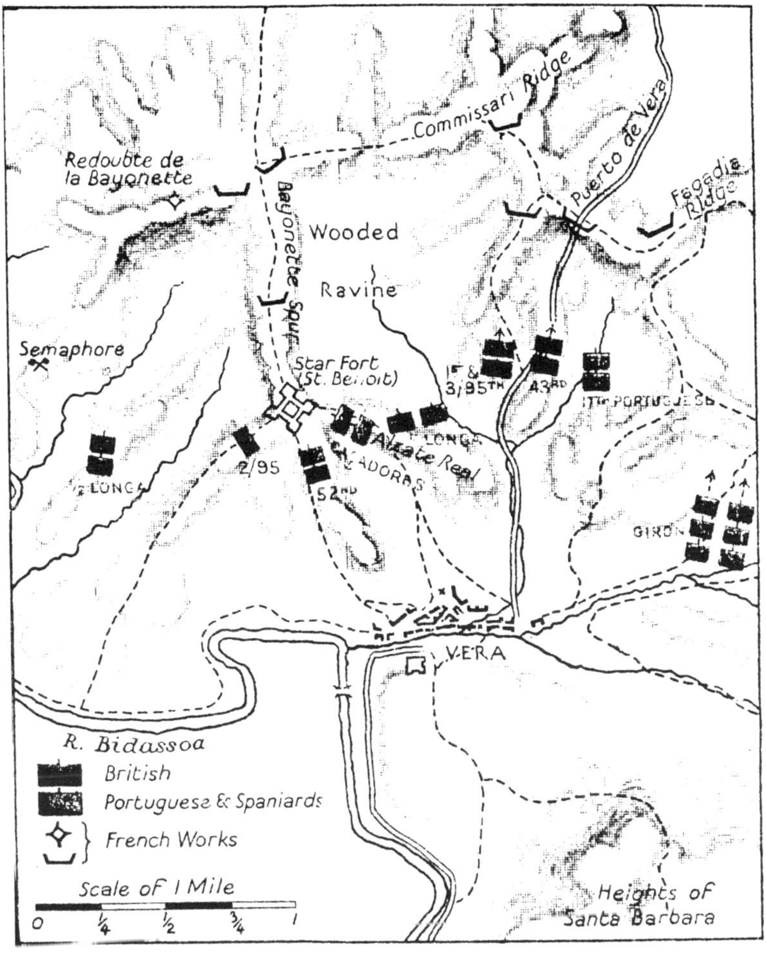

Storm of the French Lines above Vera 17 October 1813.

numbers of both Allied and French forces did not all take part – more likely Allies 6,500 and French 4,700).

The light division attacked up two spurs, Kempt's brigade on the right, and Colborne's on the left, linked by Longa's Spanish infantry; Kempt's men had a difficult uphill climb thickly strewn with rocks, briars, small trees and brush and one high prominence, the Hog's Back, but all were surmounted and eventual junction made with Colborne. He had had to take prepared defences, a stone-built redoubt and other field-works, but succeeded in driving the French from them. Giron's Spaniards

The Battlefield of Vera. Taken from Baionette Ridge, looking up at the Greater Rhune as Wellington's attacking soldiers would have seen it when advancing up this slope to the Star Fort beyond.

fought hard to take the Great Rhune without success; however, it was evacuated next day after being outflanked.

Casualties

Allies: 800 (Spanish 450) killed, wounded and missing.
French: 1,250 killed, wounded and missing.

The October combats resulted in Soult withdrawing to a line of defence on the Nivelle River running from St. Jean de Luz on the coast, along the river line, including the large natural obstacles of the Greater and Lesser Rhune, to St. Jean Pied de Port in the interior. In naturally strong positions, heavily fortified with numerous artillery redoubts, Soult had some 63,000 men to oppose Wellington's 80,000 which, however, included some doubtful Spanish units.

THE BATTLE OF THE NIVELLE – 10 NOVEMBER 1813

Opposing forces

The Duke of Wellington, commanding the Allied Army, 9 Anglo-Portuguese infantry divisions plus 3 independent brigades (29 brigades; 112 battalions – 63,143 British and Portuguese). Spanish infantry formations – 6 divisions – 25,673. 5 British batteries artillery; 3 Portuguese batteries. *Total: 88,816*

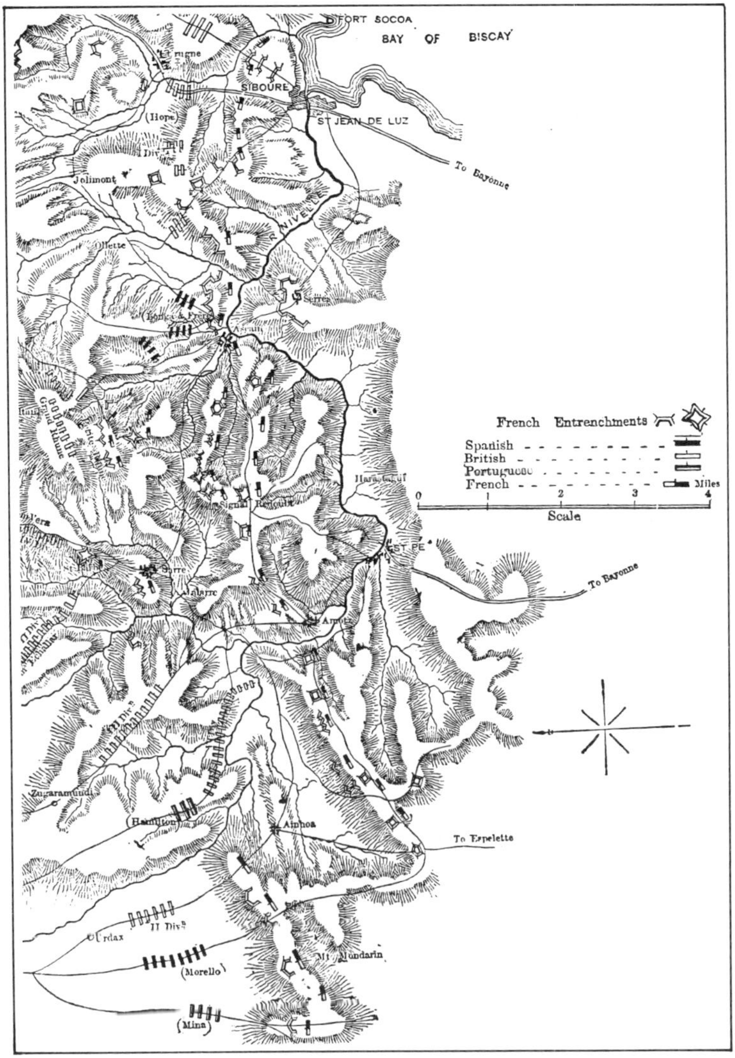

Battle of the Nivelle 10 November 1813.

Marshal Soult, commanding 10 infantry divisions (108 battalions – 57,243); 97 guns.

Anxious to be out of the Pyrenees before winter arrived, Wellington mounted a demonstration on the coast and directed his main effort of 3 divisions on each side of the river Nivelle with two more assaulting the Lesser Rhune. At daylight the Light Division, who had crept down from their eyrie on the Greater Rhune by night, assaulted the three forts on the Lesser Rhune and took them. Simultaneously, Beresford's three divisions on the right occupied Amotz. By midday, the French had been driven back north of the Nivelle, and Hill's three divisions were making good progress on the right. At dusk the Allies were in possession of St. Pee and Soult ordered a retreat to the Nive Lines near Bayonne.

Casualties

Allies – 2,700 killed, wounded and missing.
French – 4,450 men and 70 guns.

Wellington was now facing strong defences in the area between the Bay of Biscay and the rivers Adour and Nive, leading northward towards Bayonne, held by Soult with 62,000 men. Unwilling to tolerate the looting and atrocities perpetrated by Spanish troops now that they were on French soil, the Duke of Wellington sent all back to Spain with the exception of Morillo's experienced division, leaving him with about the same numbers as Soult could call upon – 63,000 men exclusive of cavalry, formed of 36 British; 23,000 Portuguese and 4,000 Spaniards. To avoid being cooped up in a triangular salient, Wellington had to get his army across the River Nive so, despite bad weather, an operation was planned for early December.

THE BATTLE OF THE NIVE – 9–13 DECEMBER 1813

Opposing forces

The Duke of Wellington commanding an Allied Army – under Lieut. General R. Hill: 1 British; 1 Portuguese; and 1 Spanish infantry division: (10 brigades; about 33 battalions – 17,380 men); 12 guns. Under Marshal Beresford 2 British infantry divisions (6 brigades; 24 battalions – 14,063); number of guns not known. Under Lieut. Gen. Sir John Hope: 3 British infantry divisions: 1 British and 2 Portuguese independent brigades (11 brigades; 28 battalions – 22,150 men); number of guns not known. In reserve: 1 British infantry division plus 2 brigades (5 brigades; 18 battalions – 10,580 men). *Total of infantry* – 64,153 men. Total artillery (including Hill's 12 guns) 5 British batteries; 3 Portuguese batteries.

Marshal Soult, commanding 8 infantry divisions and the Reserve – about 62,000 men. Number of guns not known.

On 9 December, Hill and Beresford crossed the River Nive on a front of about 5 miles; Soult, finding his 3 divisions up against 4, offered little resistance and retired towards Bayonne. Hope was sent with 3 divisions and 3 brigades to the area between Nive and the sea, where they established piquet-lines and withdrew 2 divisions, leaving the light division. Soult attacked them on the morning of the 10th, and they withdrew in the face of 4 French divisions, until stopping the French at Arcangues; the fighting was over by mid-afternoon. The same French attack

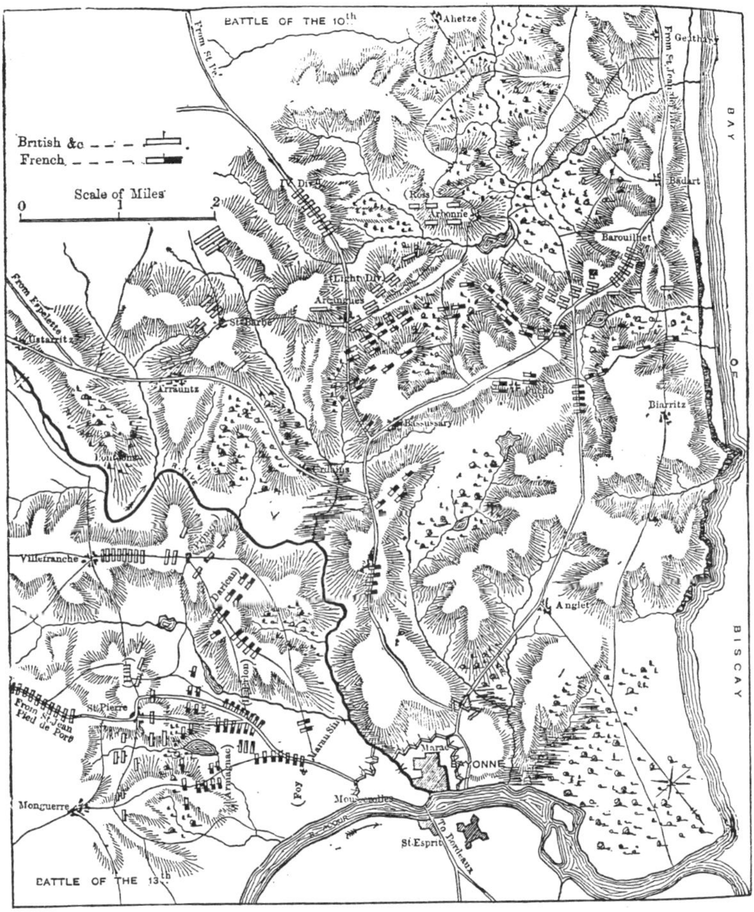

Plan of the Battles on the Nive.

The Battlefield of St. Pierre (the Nive), the Croix de Mougerre (with the Soult Monument in background). Much of the fighting took place over this area. Beyond are the hills where Ashdown and Barnes fought.

drove back the two Independent Portuguese brigades to the area of Chateau Barrouillet where, despite being reinforced, they were outnumbered 3:2; but, with fighting taking place mostly in woods, the French were repulsed. At Arcangues, the British lost 225; the French over 400; losses around Barrouillet were about even – 1,500 each. In the evening, 3 complete German battalions in French service deserted, with all their officers, colour and baggage. On the 13th, a sudden rising of the Nive caused Hill's 14,000 men and 12 guns to be cut off from the rest of the Allied army; however, fighting at odds of more than 3–1, in four hours Hill, quite unaided and after initial reverses in very heavy fighting, fought Soult to a standstill then drove him back to where he had started.

Casualties **(at St. Pierre)**
Hill 1,775 killed, wounded and missing.
Soult 3,300 killed, wounded and missing.

On 14 December, Soult moved his divisions and his H.Q. further inland, on the Adour River some 20 miles east of Bayonne; his force totalled 62,500 which included the garrison of Bayonne. In mid-February 1814, Wellington advanced and manoeuvred Soult from the Adour River line, then left Bayonne invested by a detached force and followed the French to where they had concentrated on a ridge near the town of Orthez; here they were brought to battle.

THE BATTLE OF ORTHEZ – 27 FEBRUARY 1814

Opposing forces

The Duke of Wellington, commanding the Allied Field Army. 5 British and 1 Portuguese infantry divisions; part of the light

The Battlefield of Orthez. The Church of St. Boes, prominent during the battle.

division (18 brigades; 68 battalions (43 Portuguese) – 39,157 men); 3 cavalry brigades (7 regiments – 3,373 men); 43 guns.

Total – 43,000

Marshal Soult, commanding L'Armée d'Espagne. 6 divisions (say 45 battalions – 35,000 men); 1 cavalry division (5 regiments – 1,500 men); 8 batteries artillery – 48 guns. *Total – 36,500*

With the siege force under Hope at Bayonne totalling 18,000, Wellington had 43,000 men under Hill, Beresford, Picton and himself to attack the French at Orthez, where Soult had his army positioned on a long ridge running from St. Boes village in the west, sweeping widely east and then south down to Orthez on the river. Beresford attacked and captured St. Boes, but had to pull back from Reille's spirited counter-attack. The light division were in firm control of the Roman Camp area, but the two centre divisions were in trouble until Wellington sent Colborne and the 52nd into a gap that allowed them to assail Reille's flank and rear, at the same time the Duke personally led a two-battalion

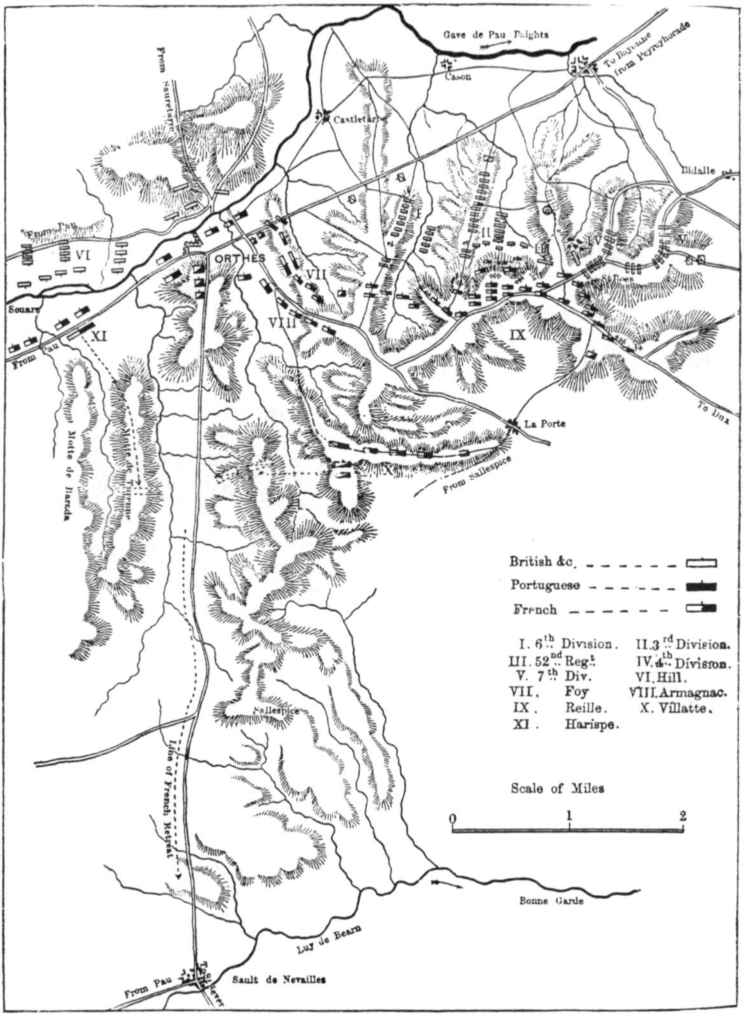

Plan of the Battles of Orthes.

attack against the French inner flank. Beresford regained St. Boes while two centre divisions cleared the French off the central plateau. With Hill's force at Orthez itself threatening Soult's left and rear, the French retreated swiftly and by nightfall had crossed the River Luy 8 miles away.

Casualties

Allies – 2,164 including Wellington himself wounded slightly.
French – 4,000 including 1,350 POWs. There were also numerous desertions.

By mid-March Wellington had driven Soult halfway across south-western France, directing the Allies towards the city of Toulouse which Soult was obviously prepared to defend. By 8 April, Wellington had his 46,000 troops (including 10,000 Spaniards under Friere) across the Garonne and closing-in on the city.

THE BATTLE OF TOULOUSE – 10 APRIL 1814

Opposing forces

The Duke of Wellington, commanding an Allied Army. 5 British infantry divisions; 1 Portuguese division; 2½ Spanish divisions (16½ brigades; about 83 battalions (29 Portuguese; about 18 Spanish) 40,706 men; Cavalry actually across the Garonne – 3 brigades (probably including Ponsonby's and Grant's) 9 regiments – 3,617 men); 46 guns. *Total say 46,000*

Marshal Soult, commanding L'Armée d'Espagne. 6 infantry divisions + 1 Reserve division of conscripts (7,000 men) (probably 52 battalions – about 36,000 men) 1 cavalry division (2 brigades) 2,700 men; Artillery – numerous (reinforced from the Arsenal at Toulouse). *Total – 42,307*

Beresford was to make an approach march of 2 miles under fire, to attack the Height of Calvinet with the 4th and 6th divisions and flanking cavalry; Friere's 2 Spanish divisions were to attack the northern edge of the ridge; 3rd and light divisions to advance on the city itself; while Hill was to make a diversionary attack against the St. Cyprien suburb. Soult had a division defending the area between the Languedoc Canal and the northern edge of Calvinet; 2 divisions on the heights themselves, with 1 in reserve; the cavalry were to hold the eastern and southern approaches; with the conscripts stationed along the ancient walls of Toulouse. Beresford succeeded after two attempts, but Friere's Spanish divisions, attacking alone at the wrong time, were driven back with heavy casualties. After bitter fighting and the 4th division brought through the ranks of the exhausted 6th, the centre of the position was secured, and the Great Redoubt at the end of the ridge evacuated as Soult pulled back into the city. On the following day, aware that the Allies held commanding heights overlooking the city, Soult retreated south to Carcassonne; the Allies entered the empty city.

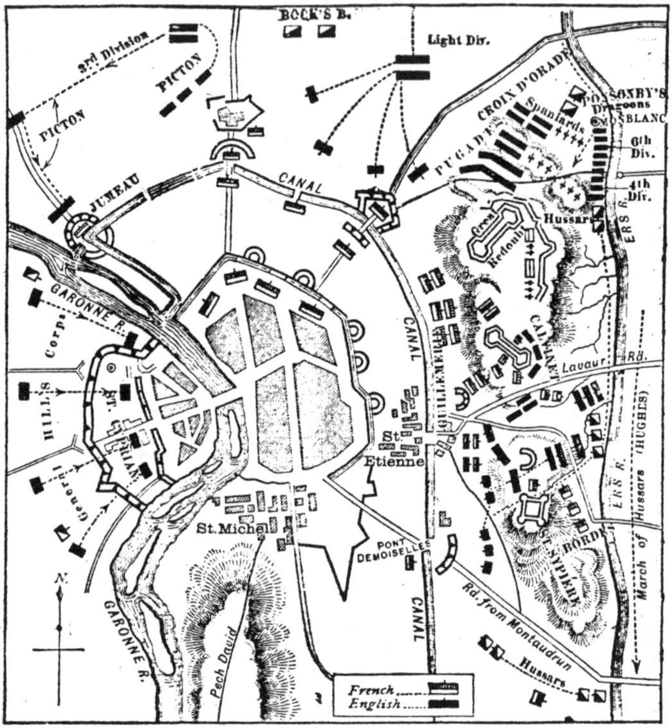

Battle of Toulouse.

Casualties

Allies – 4,500 killed and wounded, including 2,000 Spaniards (25 percent of Friere's strength)

French – 3,256 killed and wounded.

A few hours later the news arrived that Napoleon had abdicated and, after almost 6 years, the fighting was over. British cavalry and artillery proceeded across France to Calais; Anglo-Portuguese divisions marched to Bordeaux, both conscious that they had been equal partners in many hard battles.

BIBLIOGRAPHY

Books on the Peninsular War are legion – the Regimental Histories of every British Regiment (and the Royal Artillery) deal proudly and at length with their feats in Portugal and Spain during the momentous years 1808–1814. Reading any one of the books listed below will provide a reasonable knowledge of the War, to read them all would make an authority of the reader!

Brett-James, A.	*Wellington At War 1794–1815* (1961)
	Edward Costello, Peninsular & Waterloo Campaigns (edt 1967)
	Life in Wellington's Army (1972)
Bryant, Arthur	*Jackets of Green* (1972) (Collins)
Chandler, David	*The Campaigns of Napoleon* (1967) (Weidenfeld & Nicolson)
	Dictionary of the Napoleonic Wars (1979) (Arms & Armour Press)
Esdaile, Charles J.	*The Spanish Army in the Peninsular War* (1988) (Manchester University Press)
Featherstone, Donald	*Wargames Through the Ages Vol. 3 1792–1859* (1975) (Stanley Paul)
	Featherstone's Complete Wargaming (1988) (David & Charles)
Gates, David	*The Spanish Ulcer* (1986) (Allen & Unwin)
Glover, Michael	*Wellington as Military Commander* (1968) (Batsford)
	The Peninsular War (1974) (David & Charles)
	Legacy of Glory – Joseph of Spain (1973)
	Wellington's Army in the Peninsula (1977) (David & Charles)
Haythornthwaite, Philip	*Weapons and Equipment of the Napoleonic Wars* (1979) (Blandford)
Hibbert, Christopher	*Corunna* (1961)
	Recollections of Rifleman Harris (edt 1970)
	A Soldier of the 71st (edt 1976)

Kincaid, Sir John	*Adventures in the Rifle Brigade* (nd) (MacLaren)
Lawford, J.	*Wellington's Peninsular Army* (1973)
Lawford J. & Peter Young	*Wellington's Masterpiece – Salamanca* (1973)
Longford, Lady	*Wellington – the Years of the Sword* (1971)
Napier, W.	*English battles and Sieges in the Peninsula* (rpt 1973) (John Murray)
North, R.	*Soldiers of the Peninsular War* (1972)
Oman, Sir Charles	*The History of the Peninsular War* (rpt 1970) (Greenhill Press)
	Sir John Moore (1953)
	Wellington's Army (rpt 1970) (Cambridge University Press)
Richards, D. S.	*The Peninsular Veterans* (1975) (McDonald & Janes)
Rogers, H. C. B.	*Wellington's Army* (1979) (Ian Allan)
Surtees, W.	*Twenty-Five Years in the Rifle Brigade* (rpt 1973) (Longmans)
Weller, Jac	*Wellington in the Peninsula* (1962) (Kaye & Ward)
Windrow, Martin & Embleton, Gerry	*Military Dress of the Peninsular War* (1974) (Ian Allan)

INDEX